141 LESSONS FROM THE SKY

AGRICULTURAL PILOTS

FLETCHER MCKENZIE

SSP

This edition published 2025 by Squabbling Sparrows Press

ISBN 978 1 991331 54 0 (Paperback)
ISBN 978 1 991331 53 3 (Ebook)

Cover Photography:
Air Tractor 502XP: Chris Charles - Unsplash License
Grumman G-164 Ag-Cat: John Torcasio - Unsplash License
Cresco: Brent Robertson

Published by Squabbling Sparrows Press
PO Box 4213, Marewa, Napier 4143
New Zealand

Squabbling Sparrows Press

ALSO BY FLETCHER MCKENZIE

LESSONS FROM THE SKY

141 AGRICULTURAL PILOTS
STORIES & LESSONS

CONTENTS

"Ag pilots are some of the best pilots in the sky. They perform more aerobatic maneuvers in a day than most pilots do in a year."

    -Brent Murphree, Delta Farm Press

FROM THE PILOT'S SEAT

For aircraft videos, in-depth interviews with ag pilots, warbird veterans, and more, scan the QR code or visit:

www.YouTube.com/@FromThePilotsSeat

Real pilots. Real stories. Lessons that save lives.

Dedicated to the Tylor Johnson Legacy Foundation - honoring Tylor's legacy of quiet leadership, professionalism, and a commitment to safety that continues to save lives in agricultural aviation.

The Tylor Johnson Legacy Foundation honors the life of Tylor Johnson, a third-generation agricultural aviator from Minnesota who tragically lost his life in a wire-strike accident.
Tylor's family launched the foundation to help prevent future accidents by investing in what mattered most to Tylor - pilot safety and education.
One of the foundation's key contributions is its sponsorship of the "Flying in the Wire and Obstruction Environment" course at the annual NAAA Ag Aviation Expo. The full-day course, led by Utilities/Aviation Specialists Inc., trains ag pilots to recognize and avoid low-level obstacles through enhanced situational awareness, cockpit discipline, and mission planning.
As Tylor's family put it, "We can't bring him back, but we can help bring someone else home."

www.tylorjohnsonlegacyfund.com

INTRODUCTION
TOBY MCPHERSON, TALL TOWERS AVIATION

I've been flying and spraying crops for over 40 years, and if there's one thing I've learned in that time, it's that agricultural aviation isn't just a job—it's a calling. It's a life that chooses you as much as you choose it. I started Tall Towers Aviation back in 1980, naming it after the two massive broadcast towers that dominate the horizon of Traill County—symbols of connection, endurance, and perspective.

The KVLY-TV mast near Blanchard, once the tallest man-made structure in the world, stands 2,063 feet above the plains. Just a few miles away, the KRDK-TV mast climbs to 2,060 feet, making it the second-tallest structure in the U.S. These towers are visible for miles and, to me, have always represented the kind of strength and purpose we try to carry in this business. They remind us that we're rooted here—in the soil, the seasons, and the skies—and they've guided my flying more times than I can count.

In 1980, Tall Towers Aviation was a one-plane operation—like most outfits back then. Now, we're a multi-generational, family-owned business flying turbine aircraft, running a full-time crew, and serving a tight-knit rural community we're proud to be part of.

But, even now, after all these years, every time I take off, I carry the same feeling I had back when I was just a kid watching my older brother and the neighbor crop sprayers with wide eyes. By second grade, I knew I wanted to be in that cockpit.

Flying came first. But being a great crop duster is about more than just the airplane. You've got to be the agronomist, the mechanic, the meteorologist, and sometimes the weatherman, all in one. You've got to know what you're spraying, when you're spraying, and why you're doing it. You're reading the wind, watching the terrain, anticipating drift, adjusting for airspeed and payload, and calculating risk every few seconds. Every pass matters. Every load counts. And every field is different.

Back when I started, I flew a 150-horsepower aircraft that carried 100 gallons. Now I fly a 750-horsepower turbine machine with five times the capacity. The upgrades have been huge—air conditioning, GPS guidance, flow control, real-time tracking. The work's gotten faster, smoother, and safer in many ways. But one thing hasn't changed: the margin for error is still razor thin. Powerlines don't forgive complacency. Turbulence can still surprise you. Weather can change on a dime. And these days, we're sharing the skies with everything from wind turbines to drones.

But it's not just the flying that's evolved—it's the entire industry. I've always believed in staying curious and learning every day. That mindset has helped me survive some tough seasons. I've worked with and learned from some great people along the way. My brother Jerry was a mentor early on. The encouragement and mentorship of my friends Bob Odegaard and Gerry Beck, shaped a lot of who I became as a pilot. Losing both of them in accidents has been one of the hardest things I've faced. But even then, you've got to get back on the horse. That's what we do in this line of work. You get back in the saddle, honor their memory, and keep flying smarter.

I've tried to stay ahead of the curve by embracing change, especially in technology.

I've taken a strong interest in drones and precision ag systems—because I see the direction we're heading. Companies like Growings are investing millions into next-gen drones that might one day assist or even supplement some of what we do. I've even supported testing by providing GoPro footage from my own aircraft. I don't see these innovations as threats—they're tools. If used right, they'll help us be even better at what we do: apply product with precision, protect yields, and do it safely.

I'm proud to serve on the board of the National Agricultural Aviation Association (NAAA), where I've had the chance to help shape policy, improve pilot safety, and support the next generation of ag pilots. I'm also part of a team establishing the Fargo Air Museum, because preserving the legacy of this profession matters. Ag aviation has always played a vital role in food production, and we need to tell that story to the public, to the media, and especially to young people considering this path.

Because here's the truth: we need new pilots. The average age of an aerial applicator today is 54. If we don't attract fresh blood into the industry, the expertise we've built over decades risks fading out. That's why mentorship, training programs, and books like the *Lessons from the Sky* series are so important.

This book isn't just a collection of stories. It's a flight log of lessons—some painful, some powerful, all worth remembering. The good landings, the close calls, unexpected engine failures, wind shear, the wrong field sprayed—they all have something to teach. Whether you're in North Dakota, New Zealand, or Brazil, the fundamentals are the same: fly sharp, stay humble, and never stop learning.

I've had the honor of flying historic aircraft across the Caribbean, flying warbirds with the Dakota Territory Air Museum, and taking part in commemorative flights like the Arsenal of Democracy (75th World War II Victory Commemoration) over Washington, D.C. I've been recognized with environmental awards from DuPont, and named North Dakota Operator of the Year more than once. But those aren't the things I carry most proudly.

What I carry is the responsibility of passing on what I've learned to my sons, my daughter, to my crew, and to *you*, the reader, the aviator.

You can't have a bad day in this profession. I smile every day I get to fly. There's too much at stake. If you keep your edge, if you prepare for the worst and hope for the best, and if you never stop learning, then this life will reward you tenfold.

We've come a long way from 55-gallon barrels with slide gates and hand-cranked dispersal systems. But we still have a long way to go. And if we keep sharing stories like the ones in this book—if we listen, reflect, and improve—we'll make sure the next generation of pilots is even better prepared than we were.

That's what *Lessons from the Sky* is all about.

Toby McPherson
Founder, Tall Towers Aviation, Page, North Dakota, USA.
Board Member, NAAA & Fargo Air Museum

PREFACE

As a kid, I was fascinated by aircraft - thanks largely to my dad. He passed down his passion for aviation through stories, model kits, and endless discussions about Boeing versus Airbus. I remember telling my mother she obviously named me after the Fletcher aircraft. As it turns out, "Fletcher" was a family name of French origin - meaning arrow maker. A good fletcher would shape and place the fletching (feathers) to ensure an arrow flew straight and true. I've always loved that metaphor. In some ways, this book is a modern fletching - helping pilots stay on course and fly safer.

Growing up in New Zealand, I was drawn to the unmistakable silhouette of the FU24 Fletcher. Originally designed in California by John Thorpe - who also helped design the Piper Cherokee - the Fletcher became a rugged agricultural icon in New Zealand. With STOL performance and brute reliability, it transformed aerial topdressing and earned the respect of an entire generation of pilots.

As New Zealand farming scaled up, the industry moved from Tiger Moths and PL-12 Airtruks to Crescos and, eventually, to the modern SuperPAC.

I consider myself lucky to have played a role in working on marketing and selling the latest derivative of the Fletcher - bringing its legacy to a new generation of operators around the world. I asked Josh Calder, from Rural Air Work, to send a photo of his distinctive hot pink Fletcher derivative, the Cresco, for the cover.

My obsession with aviation safety didn't begin in the cockpit. It began with a high school maths project in 1990. I wanted to join the Royal New Zealand Air Force (RNZAF), so I focused my assignment on aircraft accidents in New Zealand. With no internet, I phoned the Civil Aviation Authority and ordered eleven thick journals that documented the nation's crashes. As a teenager, it was sobering. These weren't just numbers - they were real people. Pilots who didn't come home. The project planted a seed: a lifelong desire to learn from what went wrong, so others might avoid the same fate.

Years later, I read a pilot's story titled, *Seat Failure on Take-off*. It described how a seemingly minor oversight - a seat sliding back on takeoff - could become catastrophic. I investigated my own Cessna seat. I'd assumed four legs held it firmly in place. Back at the aero club, I crawled under the seat and was stunned to discover the truth: only one 5mm steel rod, a single leg, was holding my 90kg frame in place. Not four. Not even two. One. I'd flown with passengers heavier than me. I couldn't believe I had never questioned it.

So I did what many pilots never do - I researched aftermarket seat locks. Eventually, I found one for the 172 and installed it immediately. It doesn't hold the seat in place - it simply stops it from sliding all the way back. But it's enough. It's one more layer in the safety net. One less hole in the cheese.

The phrase, the *Swiss Cheese Model*, is my favorite way to think about safety. Developed by James Reason and Dante Orlandella, it illustrates how accidents happen not because of a single failure, but because multiple layers of defense all happen to break down at once. Each layer - training, checklists, habits, equipment - has holes.

And when those holes line up, the result is tragedy. But change one habit, add one safeguard, plug just one hole - and the entire trajectory of failure is blocked.

That model became real when I almost killed my daughter.

After a routine maintenance job, a grease pottle had been left loose inside the fuselage. Just before takeoff, it jammed under the aileron, rendering lateral control useless. At the point of the turn, we would have been 200 feet up - low, slow, and vulnerable - and control would have slipped into chaos. My mind spiraled. But because I did the final check I always do, "Are my controls full and free?", I realized that they weren't.

I stopped, shut down, and sat there, stunned. Not because of what happened, but because my daughter was in the glider behind me.

For months, I replayed that moment, telling myself that I'd nearly killed her. That maybe I shouldn't fly again. But eventually I saw it differently: I had saved her life. Not by luck, but by being a pilot who always follows the checklist. That moment changed everything. I committed to becoming the best pilot I could be. I immersed myself in lessons from others - because sometimes, one habit, one decision, is the only thing standing between disaster and a second chance.

Years later, it happened again - this time with my other daughter onboard. At 700 feet over water, our engine failed. Sixty seconds of silence. Sixty seconds to make it count. But this time, I wasn't paralyzed - I was prepared. Weeks earlier, I'd practiced engine-out drills, stalls, and glide performance. I knew the aircraft. I trusted the training. I flew the aircraft. And I brought us both home.

That flight reinforced what I had learned the hard way: the best chance we have is knowing the aircraft, respecting the process, and never assuming we're immune.

Complacency and fatigue are two of the most dangerous holes in the cheese. In Chapter 2 of my first book, *81 Lessons From The Sky*, there's a story about a pilot who mistook rough ground for a runway. They survived. The aircraft didn't.

Their words stayed with me: "My brain had locked into a decision, and my concentration was on the technicalities of making another smooth landing." It sounded all too familiar.

In 2009, after filming at the Classic Fighters Airshow in Omaka, New Zealand, I flew home in a Cessna 172. Clear skies. Good fuel. A straightforward VFR route - until it wasn't. Entering Ohakea's control zone, we encountered worsening cloud and turbulence. The compass spun. The ADF was useless. Still, I pressed on. When we broke through and saw a lake, I convinced myself it was Lake Taupo, New Zealand's largest lake. The map had to match what I saw. But it didn't. I'd been trying to force the facts to fit my assumptions. We were far off course.

That experience rattled me. When I got home, I debriefed with an instructor. His advice was simple: "Write it down." So I did. That story became the first of many. Writing turned to reflection. Reflection became a habit. And that habit became my first book, *81 Lessons From the Sky - General Aviation*.

This book is the ninth book I have published and, thanks to my short role working with the SuperPac750, I became more curious about the agricultural aviation sector globally.

In the United States, aerial application was born out of necessity as early as the 1920s - a direct response to growing agricultural threats. The first recorded trial took place in the summer of 1921 at McCook Field in Dayton, Ohio. French-born aeronautical engineer Étienne Dormoy, along with U.S. Army test pilot Lieutenant John Macready, conducted an experiment that would launch a global industry. Dormoy mounted a crude, hand-cranked metal hopper onto a surplus Curtiss JN-6 "Jenny" biplane and filled it with powdered lead arsenate. As Macready flew six low passes over catalpa trees infested with sphinx moth larvae, Dormoy manually dispersed the chemical from the cockpit.

The results were immediate - the pests were suppressed, and a new era of agricultural aviation began.

The choice of catalpa trees was no accident. Known for their durable, rot-resistant heartwood, they were widely used for telephone and utility poles, even railroad ties, helping support the U.S.'s rapidly expanding communications infrastructure. Protecting these trees was critical - and aviation proved it could get the job done faster and more effectively than ground-based methods.

As agriculture scaled and pests continued to threaten vital crops, aviation offered a game-changing solution. By the 1930s, aerial spraying spread across the country, especially in the American South, where boll weevils were ravaging cotton fields. The aircraft evolved too - from war-surplus biplanes to purpose-built ag machines designed for low-level precision work. Aviation had found its place - not just in the skies, but in the fields.

Few contributed more to that evolution than Leland Snow (1930–2011), widely regarded as the godfather of modern agricultural aviation. At just 21, Snow designed the S-1, the first purpose-built spray aircraft, and by 1958 he had launched the Snow Aeronautical Company in Olney, Texas. He later developed the S-2 series - over 500 of which were produced before the design became the foundation for the Rockwell Thrush. After selling his business to Rockwell in 1965, Snow went on to found Air Tractor in 1972, introducing the AT-300/301 series and, later, turbine-powered models that would define the next generation of ag aircraft.

Known for his humility and hands-on approach, Snow frequently visited operators, shared safety ideas, and championed innovations that improved performance and pilot protection. He remained dedicated to advancing the safety, reputation, and effectiveness of aerial application until his passing at the age of 80.

From its earliest moments, aerial application has been driven by innovation and shaped by risk.

And from those first hand-cranked hoppers to today's powerful workhorses, one truth has remained: necessity may have created ag aviation - but it's courage, engineering, and constant learning that keeps it alive.

In 2024, I met ag pilot, Toby McPherson, at the National Agricultural Aviation Association (NAAA) conference in Fort Worth, Texas. I had come to learn about the ag aviation sector in the U.S., I was also curious to see whether my *Lessons from the Sky* books might offer value here too. Toby and others shared their world with me - the daily challenges, the long hours, the critical thinking required just to do the job and come home safe. He wasn't alone. All across the U.S. and around the world, I began hearing stories. Some were close calls. Some were tragedies. All were deeply instructive.

We met attending the "Flying in the Wire and Obstruction Environment" course - thanks to the generous support of the Tylor Johnson Legacy Foundation. Tylor Johnson was a third-generation aerial applicator who tragically lost his life in a wire-strike accident. A foundation was established in his memory to enhance pilot safety by funding critical training, especially in wire-obstacle awareness and avoidance. The full-day course, delivered by Utilities/Aviation Specialists Inc., focused on low-level flying hazards, wire detection, and decision-making under pressure. It was one of the most practical and eye-opening safety programs I've experienced - and a clear example of how learning from the past can save lives in the future.

A later visit to Toby McPherson's operation in Page, North Dakota, was inspiring and humbling. Standing alongside him in the Tall Towers Aviation hanger, I saw firsthand at how four decades of experience, adaptation, and grit shaped one of the region's most respected ag aviation businesses.

From his early days with a single aircraft to today's turbine-powered fleet, Toby has built a legacy rooted in safety, precision, and community. As we walked the field, reviewed systems, and talked about pilot mentorship, it became clear that Toby isn't slowing down.

He's focused on the future: embracing technology, mentoring young pilots, contributing to aviation policy through the NAAA, and helping preserve ag aviation's history.

His operation is an example of what happens when passion meets purpose. And I left Page with a deeper respect for both the past and the direction this industry is heading.

While I'm not an ag pilot, I do tow gliders in a Piper Pawnee. I've flown long enough - and listened enough - to know that the risks are real. Low-level operations, fatigue, dispatch pressure, weather shifts, complex loading, unfamiliar terrain - this is flying on the edge. And too often, it's done alone. That isolation makes it harder to learn from others' mistakes. Harder to speak up. Harder to reflect.

That's where this book comes in.

In January 2025, I decided to create a resource for agricultural pilots worldwide. I began collecting, curating, and rewriting accident reports and real-life stories from ag pilots across the globe. Names and genders have been removed. As Patty Wagstaff put it, "The airplane doesn't know I'm a woman." The aircraft doesn't care how many hours you've logged or what you've flown - it just wants to fly. Preferably, not into anything.

This book is not about judgment. It's about staying alive. It's about adding one more layer of safety, one more habit, one more lesson that might keep you from becoming a statistic. We don't get better by hiding our mistakes. We get better by learning from them.

I hope, like me, you'll take something from these stories. Maybe you'll double-check your fuel. Maybe you'll pause when you feel pressured. Maybe you'll install that seat lock. Maybe you'll rethink what "just a short flight" really means. Every time we reflect, we close a hole in the cheese.

General Chuck Yeager once said, "It was my fear that made me learn everything I could about my airplane and my emergency equipment, and kept me flying respectful of my machine and always alert in the cockpit."

This book exists to honor that mindset.

Here's to straighter flying arrows and to safer skies.

Fletcher McKenzie

GLOBAL AGRICULTURAL AVIATION OVERVIEW

The U.S. leads the world's global agricultural aviation in scale, technology, and infrastructure. Yet, the sector's impact is truly international - with massive Brazilian operations, emerging European regulations, and increasing autonomy in Asia. As sustainability and precision become more important, ag aviation is evolving worldwide - leveraging turbine fleets in some regions and drone technologies in others - but one thing remains constant: its crucial role in feeding the world safely and efficiently.

Around the world, agricultural aviation reflects the unique demands and regulations of each region. In the United States, fixed-wing aircraft dominate, treating over 127 million acres annually with high-output turbine fleets. In contrast, European Union regulations have sharply restricted aerial spraying - especially near urban areas - leading to significant declines in manned ag aviation operations.

Countries like France and Germany have implemented near-total bans on fixed-wing crop dusting because of environmental and public health concerns. Meanwhile, Brazil boasts one of the world's largest manned fleets, driven by vast agricultural zones and minimal urban conflict, while China and South Korea are scaling rapidly, mostly by using agricultural drones. In fact, China had over 120,000 agricultural drones in operation as of 2023, many deployed in rice and tea production across mountainous terrain. Drones offer lower cost and access to tight or hazardous areas, but their limited payload, shorter endurance, and regulatory constraints mean they complement rather than replace manned aircraft - especially for large-scale farms.

The future of aerial application will involve a hybrid model, where manned aircraft continue to cover vast acreage efficiently, while drones take on precision tasks in sensitive or fragmented environments.

Aviation safety in agricultural flying is a constant and evolving challenge. Ag pilots operate in one of aviation's most demanding environments - flying low, fast, and often alone, with minimal margin for error. Hazards like power lines, changing weather, uneven terrain, mechanical stress, and fatigue all contribute to a higher-than-average accident rate.

In the United States, wire strikes remain a leading cause of fatal accidents, despite improved training and technology. Programs like PAASS (Professional Aerial Applicators' Support System), Operation SAFE, and wire-strike avoidance courses - many supported by organizations like the NAAA and the Tylor Johnson Legacy Foundation - are helping to drive awareness and reduce risk. GPS guidance, obstacle alert systems, and real-time telemetry have become essential tools, but safety still depends on pilot judgment, planning, and vigilance.

As drones enter the scene and operations become more complex, the need for a strong safety culture - one built on experience sharing, incident debriefs, and continuous learning - has never been more vital in keeping ag pilots alive.

Beyond the major ag aviation associations in the U.S., Brazil, Canada, Australia, and New Zealand, several regions are experiencing rapid change or face unique limitations. In China, agricultural aviation is undergoing explosive growth - primarily through drones - driven by companies like DJI and supported by government-led modernization efforts. India is following a similar path, with initiatives like Drone Shakti promoting drone spraying in hard-to-reach or small-plot farms, although traditional aerial application remains rare.

In Africa, countries like South Africa have active ag operations, particularly in crop spraying and pest control, though no formal national association exists. Meanwhile, the European Union has moved in the opposite direction: aerial spraying is heavily restricted or banned outright in many member states because of environmental regulations and population density.

Despite these challenges, innovation continues, especially in unmanned systems and hybrid technologies - ensuring that agricultural aviation, in both manned and unmanned forms, remains a vital tool for feeding the world.

Agricultural aviation is a critical, precision-driven sector that spans the globe, touching nearly every farming region.

United States:

- Around 1,560 ag aviation businesses support 3,400–3,500 pilots, operating small fleets of 2–3 aircraft each.
- Each year, U.S. operators treat approximately 127 million acres of cropland - about 28% of the nation's row crops - plus millions more in pasture, forest health, and public health missions.
- The combined aerial application fleet flies over 830,000 hours annually, performing 30–100 takeoffs and landings per aircraft per day - primarily with turbine-powered fixed-wing planes.

Global Landscape:

- Worldwide, agricultural aircraft markets are valued in the billions of dollars, with growth projected through 2030; new electric and unmanned systems are emerging - especially in Brazil, China, and parts of Europe and Asia.
- In Brazil, a single pilot might spray over a million acres annually, reflecting the massive scale of South American operations.
- While UAVs and electric ag-drones are gaining traction - pioneered in Japan, South Korea, and increasingly in parts of Africa - they remain complementary to manned systems; most global acreage is still treated by traditional ag aircraft.

Why It Matters:

- Globally, about 1.58 billion hectares of arable land are farmed - and aerial application plays a vital role, especially in regions with large-scale crops, challenging terrain, or time-sensitive pest outbreaks.
- In the U.S., ag pilots directly contribute to managing pests, boosting yields (~8% higher vs ground spraying), seeding cover crops, forestry and mosquito control, and rapid disaster response. Agriculture experts estimate this prevents the conversion of more than 27 million acres of wetlands or forests into farmland annually.

In putting together the *Lessons from the Sky: Agricultural Aviation* edition, we made a clear decision from the start: to only include incidents where no one was killed or seriously injured. This was about learning from close calls - not reliving tragedy - and shining a light on the moments where better decisions, awareness, or training made all the difference.

We're incredibly grateful to the people and organizations who made this work possible. With special mention to the following organizations:

ASRS: Aviation Safety Reporting System

Run by NASA in partnership with the FAA, ASRS is a voluntary reporting system where pilots, controllers, and aviation professionals share real incidents to help make flying safer. Their public database - and especially their monthly CALLBACK newsletter - is packed with honest, de-identified stories that capture exactly the kind of situations we wanted to highlight. Many of the stories featured in this book started with insights drawn from ASRS reports. You can explore more at asrs.arc.nasa.gov.

CAROL Database: Case Analysis and Reporting Online

The NTSB's CAROL system gave us access to official U.S. accident and incident reports. It was a powerful tool for fact-checking, understanding the bigger picture, and getting the wording and context right. We relied on it heavily to make sure each story was accurate and responsibly told. You can browse it at carol.ntsb.gov.

NAAA: National Agricultural Aviation Association

Thank you to the team at the NAAA, who helped guide us and pointed us toward valuable material. Their commitment to safety and support for this project meant a lot, and we're thankful for their help.

The NAAA represents over 1,500 aerial application businesses and over 3,400 ag pilots across the US. It advocates for pilot safety, science-based policy, and industry innovation. NAAA leads programs like PAASS and Operation SAFE, hosts the annual Ag Aviation Expo, and works with federal agencies on regulatory issues that affect aviation, environmental safety, and public health.

www.agaviation.org

AAAA: Aerial Application Association of Australia

The AAAA supports Australian ag pilots and operators across spraying, fertilizer application, firefighting, and pest control. It champions safety, compliance, and professionalism through programs like SpraySafe and represents the industry to CASA and state regulators. The AAAA also promotes sustainable aerial practices and the use of new technology in Australian agriculture.

www.aaaa.org.au

NZAAA: New Zealand Agricultural Aviation Association

A division of Aviation New Zealand, the NZAAA advocates for the country's aerial application operators, who play a key role in fertilizer spreading, pest control, and pasture management - especially in remote and rugged terrain. It works with the CAA, EPA, and WorkSafe to maintain safety and environmental standards, while supporting training and industry events.

www.aviationnz.co.nz/nzaaa

CAAA: Canadian Aerial Applicators Association

The CAAA represents Canada's aerial application industry, focusing on safety, efficiency, and environmental care. It works with Transport Canada and the PMRA to develop sound, science-based regulations and best practices. The CAAA also operates Operation S.A.F.E., promoting precision and drift control across diverse Canadian regions.

www.canadianaerialapplicators.com

. . .

SINDAG: Sindicato Nacional das Empresas de Aviação Agrícola (Brazil)

SINDAG is Brazil's national agricultural aviation association, representing the world's second-largest ag aviation fleet, with over 2,300 aircraft. A lead in industry training, research, environmental advocacy, and public engagement. Through events like the Brazilian Agricultural Aviation Congress, SINDAG promotes excellence and innovation in a country where aerial application is vital to large-scale crop production.

www.sindag.org.br

PART 1 - WIRE STRIKES / OBSTACLE STRIKES

"Mistakes are inevitable in aviation, especially when one is still learning new things. The trick is to not make the mistake that will kill you."
    –Stephen Coonts

CLIPPED BY THE POLE
AIR TRACTOR AT-602

The run was supposed to be simple. I'd done it a hundred times before - maybe more. Midday, sunny skies, light winds out of the northwest, and the air thick with summer heat. I launched from Winnie and climbed out to set up for an application pass just outside Beaumont. The field was wide open, familiar, and forgiving.

Or so I thought.

As I circled in low to establish my starting line, I swept across the edge of the field in a right bank. I knew the layout, or I believed I did. But just as I was lining up, I felt it. A sudden, dull jolt - more felt than heard. The airplane shuddered slightly but stayed airborne.

In that split second, I knew I'd hit something.

I leveled the wings and made a beeline back to the strip. The airplane felt off, but controllable. No engine issues. No hydraulic surprises. Just that sinking feeling in my chest.

Back at the airstrip, I lined up for landing. As I touched down, I knew I was in trouble. The left main gear collapsed beneath me. The airplane sagged and veered, the wing dragging across the asphalt, scraping and twisting. I brought it to a stop and shut everything down.

When I got out, I could see the damage clearly. The right wing was torn up, and the fuselage had taken a serious hit. But I was standing. I was unhurt.

Later, the investigation confirmed what I already knew: the right main landing gear had struck a power line pole during that low maneuver. That impact had compromised the gear and set the stage for the collapse on landing.

No mechanical failure. No weather issue. No surprise squall or system malfunction. It was me. I had misjudged the environment - missed a pole I should've seen. Or at the very least, planned around.

Lessons Learned:

Familiarity can dull your edge.

I knew the field. I knew the terrain. But in that moment, confidence became carelessness. Power lines and poles don't always show themselves. They blend in. And when you're focused on spraying patterns and wind drift, it's easy to let your awareness drop a notch.

That's when it happens.

From that day on, I've treated every pass like it's the first. I take a moment longer on setup. I trace the edges. I ask myself: "What's new? What's hidden? What haven't I seen yet?" Because sometimes, it's the thing you think you've accounted for that takes you down.

And when you're this low, there's no such thing as a minor mistake.

NOTES:

BLINDED BY THE FAMILIAR
UNSPECIFIED AGRICULTURAL AIRCRAFT

It was a hazy evening in Illinois, with visibility around eight miles and the sun hanging low on the horizon - just enough to blind you if you caught it wrong. I was out applying insecticide to a cornfield about 26 miles south-southwest of SQI. With 6,900 hours under my belt, including 2,700 in type, this wasn't my first time working close to obstacles. But sometimes, even familiarity can betray you.

Before beginning my runs, I surveyed the field. I flew the perimeter and spotted a set of high-tension electric lines crossing the field from northeast to southwest. I assessed the safest way to apply the spray would be to fly under the wires - something I'd done before.

The first pass went as planned. I dipped beneath the lines and treated the northern rows. Then I repositioned to begin spraying from the south edge. But this time, whether it was fatigue, shifting light, or just a misjudgment, I didn't drop low enough.

My top right wing clipped the lowest line. The wire snapped.

I immediately lowered the aircraft further to avoid further entanglement and maintained control. There was no crash, no injuries, and no major damage beyond what had already occurred.

This same power line had been struck before - back in 1968 - by another ag pilot. The low clearance, even with prior awareness, remains a significant hazard.

The FAA investigated. No action was taken against me, and the flight was deemed appropriate for the situation. Still, I know what went wrong: sun glare, haze, and a momentary lapse in visual judgment. I'd flown under those wires once that day, but it was on the second pass - fighting the sun - that I made the mistake.

Lessons Learned:

This incident reminded me that even familiar fields can become dangerous under shifting light.

I'd flown under those wires once that day. But the second time, haze and low sun conspired to hide them in plain sight. A safe first pass isn't a guarantee the next will be. Wires don't move, but visibility does.

In ag flying, sun angle can erase your margin in a heartbeat. Now, I reassess wire paths before every run, regardless of how recent the last one was. When you combine obstacles with environmental variables, repetition isn't safety - it's risk disguised as routine.

Trust your eyes, but only after giving them a fresh, honest look.

<u>NOTES:</u>

A WIRE IN THE HAZE
UNSPECIFIED AGRICULTURAL AIRCRAFT

It was one of those late summer afternoons in Michigan when the morning fog gives way to a thick, lingering haze. I'd just finished spraying three fields - about 265 acres in total - and was setting up for my final job of the day: a smaller patch about two miles away, at a 35-degree angle from my previous runs.

As usual, I flew a reconnaissance pass to check for obstacles like poles and wires. Everything looked clear at first glance. I completed an initial pass to the east, pulled out, and turned to enter from the eastern end, heading west. But as I lined up with the rows, the sun hit the haze just right - or wrong - and visibility dropped dramatically.

Just as I was flying into the field, something flickered in my vision - a glint, a shadow - then the unmistakable thump of contact. My right main gear had clipped the top wire of a power line. I watched in disbelief as the insulator dangled from the gear leg. Somehow, the aircraft kept flying.

I flew seven miles to a familiar grass strip near US-12 and landed uneventfully. After shutting down, I inspected the plane: no damage. I called the farmer immediately and asked him to notify the power company. They had service restored by 8:00 p.m. that night.

A few days later, I returned to the scene in daylight and clear conditions. That's when I saw it - a nearly invisible set of wires cutting across the northeast corner of the field, running perpendicular to the ones I had initially checked. From the air, in haze, they had blended perfectly with the ground and surrounding foliage. I'd never seen them. And that was the problem.

Lessons Learned:

This flight taught me that a single recon pass in marginal conditions isn't always enough. Wires that seemed absent from one angle were hiding in plain sight - masked by haze, terrain, and poor contrast.

Even with 2,000 hours logged, I nearly learned that lesson the hard way. From now on, I double-check wire locations from multiple directions, especially if visibility isn't perfect. Trusting a quick glance is a gamble, not a safety check.

Whether it's late-summer haze, early fog, or fading light, conditions can conceal hazards you think you know. Always give yourself margin - because what you miss might just be hanging in the air.

NOTES:

BLINDED BY THE LIGHT
UNSPECIFIED AGRICULTURAL AIRCRAFT

It was the first flight of the day - routine, or so I thought. I was spraying a 50-acre alfalfa field nestled in a narrow valley about 25 miles south of Hermiston, Oregon. The field sat beside the ranch house of the landowner, the creek-bed floor surrounded by steep hills rising 300 to 400 feet. A county road curved along the southern edge of the field, and a power line ran alongside it. I knew the terrain well - or at least I thought I did.

I had surveyed the field both from the ground the day before and from the air before the application began. Everything seemed clear.

As I made my first low pass across the field, something unexpected happened: the aircraft struck wires stretched across the eastern portion of the field. A service line had been strung from the main road-side power line to a pole near a grove of trees just outside the northwest edge of the field. I hadn't seen them.

Thankfully, the aircraft remained completely controllable. There was no damage to the airframe, no impact on ground structures - aside from the wires themselves. Power was cut to the ranch until crews could arrive a few hours later to restore it.

I was surprised, and shaken, that I missed them.

Normally, visual clues like poles, shadows, or hardware give away the presence of wires, even if the wires themselves aren't visible. But on that morning, the conditions were deceptive. The sun lit half the field in bright yellow light, while the other half was shadowed by the surrounding hills. The contrast between blinding glare and shadowed terrain made it incredibly hard to detect that stretch of wire.

Lessons Learned:

I'd flown that Oregon valley field before, surveyed it from the ground and air - but I missed one wire.

On the first spray pass, I struck a service line strung across the eastern edge. No damage, no injury, just a sharp jolt and power out to the ranch.

The culprit? A brutal mix of sunlight and shadow, masking the wire completely. Even with 9,000 hours, I was blindsided.

It taught me this: wires don't just hide in bad weather - they vanish in bad lighting too. A familiar field can still surprise you.

Always check from every angle.

And never trust what you think you saw.

NOTES:

DISTRACTION AT LOW LEVEL
UNSPECIFIED AGRICULTURAL AIRCRAFT

During an agricultural spraying operation, I was in the middle of a routine pass applying fungicide to a cornfield. As I began my third pass, I flew directly into a three-phase power line at the end of the field. Fortunately, there was only minor damage to the aircraft, and I was uninjured, though my pride certainly took a hit.

There were multiple distractions during this pass that contributed to my lack of awareness:

- The field's north end was cluttered with large trees and small sections of corn, which made it difficult to maintain accurate spray coverage, especially with the turbulence that was intensifying at noon.
- Large power lines to my right felt intimidating, contributing to my stress during the pass.
- My focus was primarily on the GPS swath tracking system to ensure the pass was properly aligned with the prescribed spray route. This led to my eyes being fixed on the monitor inside the cockpit instead of scanning for obstacles ahead.

I had become so focused on the GPS display that I lost sight of the power lines. I realized that this was a serious lapse in situational awareness, compounded by the distractions around me.

Lessons Learned:

This event underscores the dangers of human-machine interface distractions. While GPS systems and mapping tools are useful, relying too heavily on them can divert attention away from critical outside observations. I had been staring at the GPS monitor to ensure I was on track with the spray application, but this caused me to neglect my primary responsibility of scanning for obstacles, including power lines.

Going forward, I plan to be more vigilant about looking outside, especially in areas with potential hazards like power lines. I will also reassess the use of GPS systems for spray pattern monitoring, as they seem to make the task more challenging and dangerous. It may be safer and more efficient to focus on direct visual observation rather than relying on the system.

In addition, I will consider the effects of turbulence on low-level operations. Turbulence can make it harder to maintain control and accuracy, especially when flying near obstacles.

This experience also raised concerns about company policy and procedures, particularly regarding the use of technology during flight. As pilots, we need to balance efficiency with safety, and sometimes the tools meant to help us can become sources of distraction.

NOTES:

EYES AFT, TROUBLE AHEAD
UNSPECIFIED AGRICULTURAL AIRCRAFT

It was supposed to be a routine calibration flight. I had just departed Falls City (FNB) to test and verify my spray system using water over a field near my home. With clear skies and 15 miles of visibility, conditions were ideal for checking pattern coverage.

I circled the property and assessed for any obstacles. A power line ran across the field, but I had cleared it on both of my first two low passes without issue. On the third run, focused on the system's spray performance, I let my attention slip. I was looking aft - watching the swath - too long.

By the time I looked forward again, that same power line was suddenly much closer than I remembered.

I yanked back on the stick, trying to gain altitude, but it was too late. The landing gear clipped the line. I saw it snap in the mirror as the aircraft jolted slightly. From the cockpit, I could spot what looked like some minor damage to the rudder, but the aircraft was still flying clean and responsive.

I turned directly back to FNB and made a normal approach and landing.

Once down, I contacted the power company and reported the strike. Fortunately, there were no injuries and minimal aircraft damage. But it could have been worse.

Lessons Learned:

During a routine spray calibration flight, I let my focus drift - watching the swath behind instead of the path ahead. I'd already cleared the power line twice, but on the third pass, I clipped it with my gear while looking aft.

The aircraft flew normally, and I landed safely, but the near miss was a sobering reminder: no matter how experienced you are, complacency and distraction can close the margin fast.

Always fly the airplane first. System checks can wait. One glance too long in the wrong direction nearly became a disaster.

In ag flying, even the familiar can turn dangerous in a heartbeat.

NOTES:

OUT OF SIGHT INTO WIRE
AGRICULTURAL LOW-WING UNSPECIFIED AIRCRAFT

It was a clear Kansas morning - just me, my aircraft, and another field needing a pesticide run.

I was climbing into a routine spray pass, northbound across a familiar field about 10 nautical miles south of Hays (HLC). I'd already crossed the southern edge power lines without issue. But my concentration had been split. Moments earlier, I'd been troubleshooting a malfunctioning spray nozzle, coordinating via radio.

That added workload came at a cost.

I'd forgotten about a middle set of east-west wires in the field. Wires low enough to blend completely into the horizon from my approach angle. By the time I saw them, they were 100 feet away and closing fast.

Too low to dive beneath, too late to clear over.

I hauled back on the stick, but the propeller clipped all three wires. The lines whipped back, slicing gashes into the fabric of the right wing and tearing off the forward strut fairing. Miraculously, the aircraft held together. It flew smoothly enough that I was able to finish the job and return to base.

On the ground, the damage revealed itself: two slashes in the wing, a battered fairing, and several nicks in the prop. No injuries. No emergency landing. And thankfully, the only thing the downed wires served was a single oil well - which was repaired and running again that same afternoon.

Lessons Learned:

Distraction, not fatigue, nearly wrecked my flight.

A quick glance at a faulty nozzle and some radio chatter was enough to erase my mental map of a familiar field. I forgot the mid-field wires, low, subtle, and invisible until far too late.

In ag flying, there's no time to rebuild awareness once it's lost. I was lucky: the aircraft held, the job got done, and no one was hurt. But luck isn't a plan.

This flight reinforced a hard truth. Every pass demands a clear head. Break your focus, and the wires don't care how many hours you've flown. They're still there, waiting.

Reassess. Refocus. Fly like every field is your first.

<u>NOTES:</u>

POWERLESS UNDER THE WIRES
GRUMMAN G-164B

I was on my eighth spray pass over a hayfield just outside Tracy in the heat of a July afternoon. The run was typical - low and fast. I was pushing around 130 miles per hour at about 10 to 15 feet above the ground, threading underneath the high-tension lines that crisscrossed the field. This was a familiar field. I knew the lines were there and planned each pass to duck beneath them.

Then the engine quit.

No sputter, no warning - just a sudden, complete loss of power. A strange vibration hit at nearly the same moment. I couldn't tell if it came before or after the engine gave out, but it was there. And so was silence.

I didn't have altitude. With power gone, climbing wasn't an option - especially with another set of power lines just above me. I held my airspeed as best I could and flew under one more span of lines, aiming for an adjacent field where I might be able to land.

I was still too fast to touch down safely, so I tried to bleed off speed. I eased the plane into a shallow right turn, hoping to align with the raised dirt beds and glide out a landing. But I never got the chance.

The wing clipped trees lining the edge of the field. Then the fuselage dug in. The impact was violent, smashing both wings and crumpling the fuselage. But I survived. I climbed out shaken but unharmed.

When we tore into the wreckage, there were no easy answers. The engine had been rotating during impact - rotational scoring and metal spray deposits proved that. The fuel solenoid valve was found closed, which could've shut off fuel to the engine, but it may have shifted during the crash itself. The propeller showed signs of low power at impact, but the blades were intact and the mechanism worked. The fuel system had no blockages, the pump was functional, and though the fuel control unit was slightly out of spec, the manufacturer confirmed it wouldn't have caused a power loss.

So the answer? Unknown. All I knew for sure was that at a critical moment, my engine quit, and I had nothing but gravity and instinct to carry me down.

Lessons Learned:

Sometimes, you'll never get the "why."

In ag flying, we prepare for the risks we understand - wires, terrain, gusts, fatigue. But when a system fails without warning or explanation, you're left flying blind. What saved me wasn't knowledge of the failure - it was training, reaction time, and knowing the field.

I had no margin for error that day. No height, no time, and no backup. The engine quit, and all I had was the glide and the ground.

So now, I fly every pass like it might be the last. I rehearse emergencies in my head. I keep one eye on the exit route, one hand near the dump lever, and I always assume the engine might stop.

Because one day, it might. And if it does, there won't be time for questions - only action.

NOTES:

SNAP OF THE WIRE
AIR TRACTOR AT-502

Late afternoon sun glared off the windshield as I made another low pass over a wheat field just outside East Grand Forks. The Air Tractor AT-502 was steady beneath me, slicing through the air with purpose. I'd flown this route before - several times, in fact - and the terrain was as familiar as the back of my hand. That's probably what made me careless.

I was focused on the job, applying chemical across the swaying amber crop, when it happened. A sudden and violent jolt.

The vertical stabilizer slammed into something invisible, and the tail kicked hard. I knew instantly what I'd hit: wires. The top strand of a power line, just high enough to escape my vision, had caught the vertical stabilizer and ripped through the rudder. The airplane was still flying, but I had no rudder control.

I pulled up gently, trying not to aggravate the damage, and began scanning for a safe place to set down. I spotted a nearby wheat field - flat and forgiving enough - and turned toward it, hoping the tail would hold.

The landing was rough, but I kept it upright. As the aircraft rolled to a stop, I cut the power and sat there in the sudden quiet.

The rudder had been jammed, bent by the force of the impact. I climbed out, uninjured but shaken, and surveyed the damage. The vertical stabilizer was a mess, twisted and torn. But it could've been worse. A lot worse.

There had been no mechanical failure, no warning signs. The problem wasn't the plane - it was me. I'd failed to maintain visual separation from the wires. And in ag flying, that's often the only line between routine and wreckage.

Lessons Learned:

Wires are the invisible enemy - and they demand respect every single time.

No matter how many times you fly a field, complacency can creep in. You start to trust memory over sight. You stop scanning. And then one day, a wire you thought you knew - one you thought you were above - rips through your aircraft like a blade.

Now, I double my checks. I recon every field like it's my first time flying it. I use GPS overlays, fly higher survey passes, and never assume the wires are where I left them. Because even one foot too low can cost you the tail - and maybe your life.

In agricultural flying, it's not just the spray that matters . It's what you don't see that can kill you.

And that's a lesson written in wire.

NOTES:

SPIRALED BY THE INVISIBLE
AIR TRACTOR AT-402B

It was early morning in July - prime time for ag flying. The sun had just cleared the Iowa horizon, casting long shadows across the field I was working. The air was calm and cool, and my AT-402B was flying as smoothly as a tool well-matched to its task. This was my fourth pass of the day, and the job had been going flawlessly.

I lined up, low and steady, ready to lace the rows with the final application of product. My eyes swept the terrain ahead - no obstacles, no wires, no terrain shifts. I was locked in.

And then, everything changed.

Without warning, the aircraft jolted, like I'd hit a wall of air going the wrong way. A dust devil. Small, tight, invisible from the cockpit until it hit. A swirling column of rotating air that had built up over the warming earth and decided to drift right through my flight path.

I lost lift instantly. The airplane dropped just enough to put me into the wires.

The snap of impact was immediate. I felt the jolt in my seat before I realized what had happened. The rudder tore away from the vertical stabilizer, and the aircraft yawed hard left. With directional control gone and the aircraft veering, I chopped the power.

There was no saving the pass now - I just needed to get her on the ground in one piece.

I aimed for the same field I was spraying. The landing was rough - no rudder, no left gear. As we touched down, the left main collapsed and ripped away from the fuselage. The aircraft slid, twisted, and finally came to rest in the dirt, the nose pointing nowhere near where I'd intended.

The left wing was wrecked. The tail was broken. I was lucky to walk away without a scratch.

There were no mechanical failures, no warning signs. The aircraft had been inspected recently. It wasn't the airplane - it was the sky.

I later reviewed every decision, every movement. But no change in technique or throttle would have changed the moment that dust devil hit. It was random, invisible, and unforgiving. And it reminded me just how alive the air can be - especially close to the ground.

Lessons Learned:

Flying low demands complete awareness - but sometimes, even perfect focus isn't enough.

Dust devils are rarely discussed, barely mentioned in official training, and yet they're real hazards. They can be invisible until the moment they twist your lift away, and at ag-flying altitudes, there's zero margin.

This accident taught me that not all dangers in aviation are mechanical or obvious. Some are environmental, unpredictable, and nearly impossible to plan for. When you're flying low, you're not just flying the machine - you're flying the invisible air it's riding on.

Respect it.

Because sometimes, the clear sky hides the wildest things.

NOTES:

TANGLED ON TAKEOFF
UNSPECIFIED AGRICULTURAL AIRCRAFT

It was a typical late-spring evening over Walla Walla, Washington. I had flown fields like this before - tight, bounded, and scattered with the kind of distractions that test the most experienced ag pilot.

This one seemed straightforward enough. After surveying the area, I committed to the first spray pass. But as I pulled through the end of the field, something critical happened: I didn't climb out quickly enough.

My attention had been pulled toward a wire that ran at an angle along the edge of the field and a nearby road where I was keeping an eye out for vehicle traffic. What I missed - until it was too late - was another power line stretched across the departure end.

I hit it.

Despite the strike, the aircraft stayed airborne. I immediately contacted Walla Walla Tower and asked them to inform my company and have the power company notified. Within minutes, the fire department was on scene, diverting traffic around the downed wire.

I should have burned off the load over a more open part of the field. And saved the tighter sections for when the aircraft was lighter.

I let a combination of complacency and distraction lead me into a confined space I couldn't safely climb out of.

Lessons Learned:

This incident was a powerful reminder that the first pass can be the most dangerous. Especially when fully loaded.

In tight fields with complex surroundings, distractions multiply. A wire that blends into the background or a car near the road can easily steal your focus.

I learned the hard way that prioritizing a clean climb out path over efficiency is essential.

In the future, I'll leave the cluttered corners and narrow exits for later, when the aircraft is lighter and more responsive.

No matter how routine a job may seem, *every* field deserves a fresh look, a deliberate plan, and a healthy respect for the hazards hiding in plain sight.

NOTES:

THE GUST THAT GOT ME
ROBINSON R44 II

It was a humid August morning in Iowa, and I was deep into the rhythm of the spray season. I'd lifted off earlier that day from a field near Maynard, heading out on an application run in the R44. The helicopter had been running strong - no mechanical issues, clean inspections, and nearly new off its last 100-hour. The wind was light but gusty. The kind of day where you keep one hand ready for a correction.

As I maneuvered low over the crop, I was focused on precision. The lines were tight, the coverage even, and I was threading the machine just a few feet above the canopy. Then it happened.

A gust slammed the nose down - hard.

It was as if someone had slapped the helicopter out of the sky.

I yanked the collective and tried to recover altitude, but I was already too low. The tail swung wide and low, and before I could correct, the unmistakable sound of metal snagging echoed through the cockpit.

We'd hit something.

I managed to get the helicopter down safely, setting it on its skids.

No fire, no injuries - but when I stepped out and looked back, I saw the damage: the tail boom was bent and twisted. We'd struck a power line, one I hadn't seen in time.

The post-flight inspection confirmed what I already knew: the helicopter had no mechanical faults. The tail rotor, controls, and engine were all functioning. But the gust had taken control away from me in that one critical second.

The wire? It had always been there. I simply hadn't seen it - and the wind did the rest.

Lessons Learned:

In ag flying, wind isn't just turbulence - it's terrain.

We often prepare for obstacles we can see: fences, trees, poles. But we forget that wind - especially low-level gusts - can act just like terrain, shifting your aircraft's position with brutal force and zero warning.

That day, the gust didn't just push me it decided my altitude. And the line I'd missed became the price of my assumption.

Now, every time I fly low, I remind myself that environmental factors aren't just background noise - they're active variables. Wind can become your enemy in a blink, especially when operating just feet off the ground.

And as for wires? If I don't see them twice, I don't go in. Because in this business, what you don't see, or what you think you've already accounted for, can take your tail off - literally.

NOTES:

THE INVISIBLE WIRE
ROBINSON R44 II

It was late in the afternoon when I climbed into the cockpit of the Robinson R44 II, ready for another round of aerial spraying. The skies were clear, visibility was excellent, and the job was familiar - low-level application runs across the Imperial Valley. I'd done this work countless times before. Precision flying, tight to the terrain, quick passes over fields, managing drift, keeping speed steady. This was my rhythm.

The light was still decent. The wind, calm. I was almost done for the day.

Everything was going smoothly. I was lined up over a new field, flying my pattern low and fast, hugging the crop lines and making sure the coverage was even. I wasn't worried - there were no reported obstacles, and I'd flown this area before.

Then, just as I descended into another pass, I felt a subtle vibration. It wasn't sharp or jarring. Just enough to make my senses go alert.

In a helicopter, any vibration you don't expect is a red flag. Years of experience had taught me that. I didn't hesitate.

I pulled up and aborted the run, heading for an open spot nearby where I could set it down and take a look.

The machine handled fine. Everything felt responsive. But that vibration - it stuck with me.

Once I landed and shut down, I stepped out to inspect the aircraft. That's when I saw that one of the main rotor blades had taken a hit.

A wire strike.

I'd flown right through an unmarked power line during descent. The blade had cut clean through, but not without damage. The line had been invisible in the light. No markers. No warning. Just the perfect combination of angles and shadows to hide it from view.

The rotor blade was torn up enough to be classed as substantial damage.

The rest of the aircraft? No issues. No engine problems. Controls, hydraulics - everything checked out. The only failure had nothing to do with the aircraft itself. It was the line I never saw.

The aircraft was pulled from service for repairs. That rotor damage was too significant to ignore. Thankfully, I was fine. The landing was smooth. No other property was touched. It could've been much worse.

There was no need for anyone to come inspect the site in person. My report, aircraft records, and the flight conditions told the story well enough.

Lessons Learned:

This incident reminded me of something we all know but sometimes forget: wires are one of the biggest unseen threats in low-level flying.

You don't have to be reckless to hit one. You just have to not see it. And even the most familiar field can hide something new.

Here's what I took from it:

- Wires are often invisible, even in good light. Never assume you'll see them in time.
- Flying the same ground before doesn't mean you know every hazard.
- A quick recon pass - even over known fields - can prevent a surprise like this.
- Pre-flight planning should always include wire checks, satellite views, and landowner notes.

If something feels off, stop the run and regroup. That vibration was subtle, but it was enough. I got lucky. The helicopter stayed airborne long enough to put it down safely.

The wire didn't offer me a warning. It never does. That's the nature of this job.

In ag flying, the dangers aren't always in the weather or the machine. Sometimes, they're stretched invisibly across the very field you think you know best.

NOTES:

THE LINE I DIDN'T SEE
AYRES S2R-G6

It was a clear, warm May day in Kansas - nothing unusual. I'd taken off from Ulysses around midday in the Ayres S2R-G6, ready to treat a pasture just a few minutes' flight from the airport. Light wind, good visibility, familiar terrain. One of those flights where everything feels predictable.

But that's exactly when the unexpected finds you.

I'd flown ag missions for decades. Over 17,000 hours in the seat, with thousands of those in this exact aircraft type. I knew what to look for. Or so I thought.

I dropped low over the pasture, lining up for a pass. Eyes ahead, scanning for movement, obstacles, livestock, anything that could pose a problem. Everything looked good. I was steady, smooth, and committed to the run.

Then, without warning, I felt it. A jolt. A snag. The airplane shuddered.

I'd hit something.

The aircraft bucked slightly but stayed controllable. I eased out of the pass and climbed away. No alarms. No immediate control issues. But something didn't feel right.

Back on the ground, I found the damage - substantial. The vertical stabilizer and rudder were torn and twisted. That's when I learned what I'd hit: a zip line cable. It had stretched across the area unnoticed, thin and near-invisible, cutting through the sky like a trap.

There had been no mechanical issues, no weather challenges, no distractions. Just a missed line - one I hadn't seen, one I hadn't anticipated. And it had been enough to cripple a perfectly good airplane.

Lessons Learned:

The most dangerous obstacles in ag flying are often the ones you don't expect - and don't see.

Cables, wires, zip lines - they don't announce themselves. They don't wave flags or flash lights. They hide in the terrain, camouflaged against the background, waiting to turn a routine pass into a wreck.

Experience can make you confident. But confidence can blind you to the quiet threats. After this flight, I learned that even after tens of thousands of hours, I have to assume there's always something I've missed.

Because the cable doesn't care how many hours you've flown - it only cares if you see it. And if you don't, the sky will remind you in steel and silence.

Look twice. Fly like every wire is invisible. Because one day, it will be

NOTES:

THE MOMENT YOU LOOK AWAY
BELL 206B JETRANGER

It was a clear morning over the fields outside Cuba City. By 11 a.m., visibility stretched for miles. Light wind, blue sky, and warm summer air - the kind of conditions that make ag flying feel like second nature. I was deep into a Part 137 spraying mission in my Bell 206B JetRanger, weaving through tight passes, eyes forward, every movement precise.

I had over 2,300 hours behind me, nearly 300 of them in this helicopter. I knew its feel, its rhythm. The job was to move quickly, hit the passes clean, avoid the obstacles, and get it done safely. The margins were tight, but I was used to that.

Then, during one of the final passes, something shifted.

Out of the corner of my eye, I spotted movement on the ground. A vehicle - unexpected, close to the edge of the field - rolled into view. It wasn't where it should've been. I glanced toward it, just for a moment, trying to figure out if it was headed toward the spray zone.

That's all it took.

By the time I looked back to the nose, the wire was already there. Thin, quiet, and directly in front of me.

There was no time to react. The main rotor struck it with a sudden, violent jolt. The helicopter bucked and twisted. I lowered the collective instinctively, trying to dump lift and soften the impact, but I was already descending. The machine had lost stability.

The skids hit the ground first - unevenly. We pitched forward, then over. I remember the roll to the left, the noise, the sensation of coming apart. When we stopped moving, we were on our side in the middle of the field.

The rotor system was destroyed. The tail boom had separated entirely. Pieces of the aircraft were scattered around the crash site.

And somehow - I was alive.

Scraped and shaken, I crawled out through the wreckage. No fire, no fuel leak, no one else involved. It was over in seconds.

Later, I told the investigators exactly what happened. The helicopter had been running normally - no warning lights, no loss of power. Everything was functioning until the moment we hit the wire. There hadn't been a system failure.

A moment of distraction was the failure.

The investigation confirmed it: a wire strike during low-level maneuvering. The probable cause? My failure to maintain clearance due to distraction - specifically, shifting focus to an unexpected vehicle on the ground.

The conditions were visual and favorable. The machine was airworthy. But a split-second glance away turned a routine pass into an accident.

Lessons Learned:

Distraction is the invisible hazard in low-level flying.

In ag helicopter work, we fly just feet above a maze of threats - wires, poles, fences, tree lines. The aircraft doesn't forgive a lapse, and the ground certainly doesn't. Unlike airplanes, we don't have a runway to align with. We move in every direction, and we rely on constant, unbroken awareness.

This was a simple reminder of a brutal truth: even experienced pilots can lose the aircraft in a single second of inattention.

So here's what I've learned:

- When flying low, fly the aircraft - nothing else.
- Don't let ground movement pull your attention unless it's an immediate threat.
- Before each pass, mentally mark wire locations. Create checkpoints. Say them out loud if you have to. "Wire left, pole right, fence straight ahead."
- Use every tool - field maps, GPS overlays, preflight surveys. Don't trust memory. Build procedures that force you to stay focused.
- And if something draws your eye, abort the pass and reset. Wires won't wait for you to look forward again.

I walked away. That matters. But the helicopter didn't. A moment of distraction left it torn apart in a quiet field.

It wasn't weather. It wasn't a mechanical fault. It wasn't even a wire I didn't know about. It was one I didn't see - because I looked away.

In this business, that's all it takes.

<u>NOTES:</u>

THE OBJECT I DIDN'T SEE
ROCKWELL INTERNATIONAL S2R-G6

It was December in South Texas, and the morning light was sharp and clear. Perfect visibility. Just another day of aerial application over familiar fields. I was flying a solid old Rockwell S2R-G6 - rugged, dependable, and ideal for the low passes ag work demands. Everything on the airplane checked out. No snags. No warnings.

I was deep into a run, flying low and fast, releasing product as I skimmed the treetops and terrain. Nothing out of the ordinary - until the sudden jolt.

It wasn't a bird. It wasn't turbulence. It was the kind of impact that leaves your gut behind. The airplane lurched sideways for a split second, then steadied. No warnings, no system failures - just that one, sickening moment.

I made it back to base and landed cleanly. But when I climbed down and walked around the aircraft, the damage was clear: the left wing had taken a hit. A bad one. Something had torn through it.

A short search led to the culprit - an electrical box, dislodged and battered, sitting near the field's edge. Yellow paint streaked across its side matched the exact color of my wing. It hadn't just been a near miss - it was a direct hit.

The team checked the airplane from top to tail. Structurally, it had held up better than expected, but the damage was still substantial. The electrical box? Likely once mounted to a pole or post along the field boundary. There'd been no warning signs, no marking tape, and I hadn't seen it - no matter how many times I replayed the pass in my mind.

There were no mechanical issues to blame. No loss of control, no system fault. Just a moment of missed clearance in an environment that leaves zero room for error.

Lessons Learned:

In low-level flight, the margin between safe and wrecked is sometimes less than a wing's width.

When you're focused on the swath, monitoring spray, adjusting for drift, and maintaining speed and line - all it takes is one unmarked obstacle, hidden in plain sight, to turn routine into risk.

I didn't see the box. I wasn't expecting anything there. And that's the point.

In ag flying, familiarity can breed blindness. You think you know the field. You think you've seen it all. But every time you go low, you're flying into a space that changes - fences move, boxes appear, poles shift. And if you're not scanning like it's your first time every time, you're not scanning enough.

Now, I make a point of walking or scouting every field perimeter - especially on the first run. And I ask the crews about obstacles. Even when I think I know the answer. Because I've learned the hard way that what you don't see can take your wing - and your day - off in an instant.

Keep your eyes wide. Because the ground isn't just something to fly over - it's where the surprises live.

NOTES:

THE POWER OF PRESSURE
UNSPECIFIED AGRICULTURAL HELICOPTER

It had rained hard the day before, and the pressure was on. As a seasoned ag pilot with more than 26,000 hours under my belt, I was tasked with drying cherry orchards in Washington State before the saturated fruit spoiled. The stakes were high, and time was against us.

My morning started with a hiccup - the helicopter wouldn't start. A fellow pilot helped with a GPU boost, and I was finally airborne, 30 minutes behind schedule. I didn't realize it then, but that delay would compound the stress and cloud my judgment later in the day.

I had five fields to dry, totaling 32 acres, each packed with hazards: power lines, telephone poles, guy wires, and frost fan towers. Still, the mission was clear - dry the fields fast and thoroughly. I worked my way through the list: field #13 (2 acres), then #8 (5 acres), back to #13 (8 acres), then #5 (7 acres), and finally, the largest - field #7 with 10 acres to go.

During my high recon of field #7, I noticed some workers on the ground at a previous field. No cause for concern. I pressed on. As I was finishing the last few rows of field #7, I saw it - a power line on the ground behind me. I hadn't felt anything. No jolt. No vibration. No sound. But there it was.

My gut said, "Land now." But I was over cherry trees, surrounded by obstacles, and just a few minutes from the airport. The rotor system felt normal. The tail rotor seemed fine. I made a judgment call: finish the job, then head back and assess.

While flying to the airport, I saw another downed power line near the same area where the workers had been. I landed near the fuel pump to inspect the rotor blades. One main blade had a slight dent on the tip cap, the other looked normal. I called my boss and sent him photos. He said it looked minor. Moments later, the general contractor grounded the aircraft for further inspection.

Looking back, I know I let the urgency of the mission override my training and instincts. The need to deliver for the client clouded the more critical need: safety.

Lessons Learned:

Just because you don't feel damage doesn't mean it isn't there.

In high-pressure situations - especially ag operations filled with obstacles - the temptation to finish the job can override better judgment.

That day, I let urgency cloud my instincts. I should've landed the moment I saw the downed power line. Continuing to fly, even with no obvious issues, risked everything. The safe choice isn't always the convenient one, but it's always the right one. Experience isn't immunity from mistakes - it's a responsibility to recognize risk early and act on it.

I've learned to respect the signs, trust my gut, and prioritize safety above schedule or pressure.

If in doubt: stop, land, and live to fly another day.

NOTES:

THE SUN AT YOUR BACK
BELL 206B JETRANGER

It was late in the day over the cornfields near Cornland. The air was warm, the skies clear, and the light had turned golden - beautiful, but blinding. I was flying a Bell 206B, working through the final stages of a long day of low-altitude spraying under Part 137. The helicopter had held up well, and so had I. With over 3,000 hours in helicopters - just over 300 in this one - I was no stranger to the flow and fatigue of a full spray day.

As the sun dropped toward the horizon, I lined up for one more pass. The cornfield stretched out beneath me. The air was calm. No turbulence, no weather, no haze - just the fading heat of evening and the kind of light that photographers dream about. But for pilots, especially those of us flying low, that kind of light comes with risk.

Golden hour flattens depth, floods the windscreen, and washes out every contrast line you count on.

I descended into position, skimming the crop tops, lining up for one last run. Then it happened.

The sun was low - directly ahead - and everything in front of me became a blur of glare and haze.

For a moment, I couldn't pick out features. Just golden light, filling the cockpit, reflecting off the windscreen. I squinted, adjusted, tried to hold the line.

And in that moment, the wire appeared. Straight ahead.

Too late to react.

The main rotor mast hit it full speed. The aircraft jolted. I tried to regain control, instinctively lowering the collective and adjusting inputs, but the wire had already done the damage. I was descending - fast. The rotor system was compromised. I couldn't recover.

We hit hard in the cornfield.

The impact crushed the skids and fuselage. The main rotor shattered. The tail boom twisted off. The tail rotor and blades tore into the ground, scattering debris. The Bell 206B was finished.

But I wasn't.

I unbuckled and climbed out, bruised but intact. No fire. No secondary damage. Just the wreckage of a machine that had done its job for nearly 50 years - now resting broken in a quiet Illinois field.

Afterward, I told investigators exactly what happened. I didn't see the wire. The sun had washed it out. There had been no mechanical failure, no warning, no system issue. The helicopter was flying fine - until the wire ended the flight.

They confirmed it all. The aircraft had been operating normally. The weather was perfect. The only contributing factor was visibility - more precisely, the lack of it caused by the angle of the sun.

The conclusion was clear: I hit a wire I couldn't see, because of the light.

Lessons Learned:

Wires are dangerous. When the sun is low, they become invisible.

Experience can't stop you from missing what your eyes physically can't detect. Neither can instinct or reaction time. In the right light - or the wrong one - a clear field can hide the deadliest threat we face down low.

So what's the takeaway?

- If the sun's in front of you, treat the path ahead like it's hiding a wire.
- Build your spray plans around sun position whenever possible. If you can, don't end the day flying west.
- Use visors, glare shields, and polarization - but don't over trust them. They help, but they don't solve the problem.
- Scout the field thoroughly. Walk it. Fly above it. Use a drone if needed. Know where the wires are. Log them. Mark them. Memorize them.
- And if your visibility is compromised - don't go. Turn, climb, or abort the pass.

I walked away from this. That helicopter didn't. And while I was lucky, luck isn't a strategy. Not when the margin between spray height and a collision is measured in feet - and seconds.

So here's what I learned: never trust golden hour. It may look beautiful, but in ag flying, beauty can blind you.

See the wire before it sees you. Or don't fly at all.

NOTES:

THE TREE IN THE FIELD
AIR TRACTOR AT-502B

It was a calm morning in August, and I was out flying a routine spray job over a soybean field near Melville. I'd been up early, already a few passes in, the Air Tractor humming steadily beneath me. With 13,000 hours logged and more than half of those in this make and model, I knew the machine well - its rhythms, its limits, its quirks.

The field looked straightforward. Open, flat, and with good visibility. I was dialed in, focused on my spray pattern, making another low pass.

Then came the bang.

A violent jolt rocked the aircraft. My vision went black for a moment, a flash of silence swallowing everything.

When I came to, the plane was on the ground. Crumpled. Wings torn. Engine twisted from the mount. My head was pounding, but I was alive - belted in and lucky.

I'd hit a tree.

A single tree, in the middle of the field.

I never saw it.

Not once.

Whether it was because of the angle of the sun, a blind spot in the approach, or just the tunnel vision that comes from hyper-focus on the swath, I couldn't say. But it was there. And I had flown right into it.

Post-crash inspection confirmed what I already suspected - no mechanical issues. The airplane had been in top condition. This one was on me.

Lessons Learned:

You can't avoid what you don't look for.

Even after thousands of hours, even with experience on your side, you can still miss the obvious. Because when you're flying low, dodging towers, poles, and terrain, it's not just about controlling the airplane - it's about controlling your focus.

That tree was always there. It didn't move. But I failed to treat the field like new airspace. I assumed it was clean. I let routine dull my awareness.

Now, every time I start a run, I circle the field. I ask myself - what doesn't belong? What might I have missed? Because the day you stop checking is the day you start gambling. And in ag flying, the odds are stacked low and tight.

A hidden obstacle at 10 feet AGL is all it takes to bring everything to a halt.

From now on, every field gets fresh eyes - because even one tree is enough to end the day.

NOTES:

THE WIRE BELOW
HUGHES 369D

It was a warm July afternoon, the kind where the sun sits high and the shadows play tricks with your eyes. I was flying a Hughes 369D helicopter, working an ag job near Atchison. The air was smooth, visibility was excellent, and I was on another pass over a familiar field.

Before starting the run, I did a survey pass, noting a high-tension line along the edge of the field. No surprises - something I'd seen on jobs like this plenty of times. I logged it mentally and set up my route accordingly. What I didn't see, what I didn't expect, was what lay below it.

On my application pass, focused and low, I suddenly caught sight of another wire - lower, thinner, hidden beneath the glare and terrain. It was too late.

I pulled back, trying to duck under it. The rotor cleared, but not the skids or tail. The wire caught, and in a snap, we were going down. The Hughes hit the terrain hard. The rotors shattered, the tail boom twisted, the tail rotor wrecked.

Somehow, I wasn't hurt. Shaken, but standing. The helicopter was a different story.

After the dust settled and I replayed it all, it became clear. The wire wasn't new - it had been there. I just hadn't seen it. Or maybe I had, but only partially. The mistake wasn't just in what I didn't spot - it was in what I assumed.

We train to scan, to plan, to assess. But wires are sneaky. They blend into backgrounds, they hide behind trees, or sag between towers. One line above can distract you from another line below. And when you're flying close to the ground in a helicopter, there's no time for second chances.

Lessons Learned:

It's not always the wires you see that get you - it's the ones you think aren't there.

In low-level flying, obstacle clearance isn't optional, and assumption is the deadliest habit. Just because you've spotted one line doesn't mean the danger's over. In fact, that's often when it begins. Power lines are rarely alone, and their smaller siblings - the service lines, the drop wires - are the ones that vanish until it's too late.

Every pass demands complete awareness - not just of the obvious, but of the hidden.

From now on, I remind myself before every run: If you see one wire, expect another. And if you don't see any - look again.

<u>NOTES:</u>

THE WIRE I DIDN'T SEE COMING

BELL 206B

It was early July, and I was in the Bell 206B, making what should have been a straightforward application pass over a cornfield near Chapman, Kansas. I'd already reviewed the job with my ground crew. We talked about the hazards - power lines on three sides, plus a large feeder line heading into a nearby substation. I knew the area. I was prepared.

Before starting, I made a full 360° survey around the field. I scanned for wires, guy lines, trees - anything that could reach out and ruin your day. Everything looked clear. The middle of the field seemed open. I picked my path and dropped in.

I wasn't rushing. I was focused. The helicopter was stable, my line was good, and then out of nowhere, I felt a jolt. A sudden, sickening resistance.

I'd hit something.

The rotor kicked, the controls lurched, and I fought to stay upright. But I couldn't save it. The helicopter dropped into the corn, rolled to the left, and slammed into the field. We came to rest in the stalks, the tail boom twisted, the fuselage crumpled, rotor blades splintered.

Later, when I walked the field, I saw the wire - about a quarter-inch thick and green. It had been invisible against the background of the crops. Practically camouflaged. It wasn't part of the big feeder system. It was lower, less obvious, a smaller hazard that slipped right past me.

There had been no mechanical failure. The aircraft was performing perfectly. The only failure was mine - failing to spot that one wire that didn't show itself until it struck.

Lessons Learned:

Even the most thorough survey can miss what the eyes aren't tuned to see.

In ag flying, the enemy isn't always a tree line or a known transmission path - it's the obscure, the forgotten, the wire that blends with the sky or the crop line. A green wire against a green field is invisible until it isn't.

Preparation helps, but assumptions kill. I assumed that because I had looked, I had seen. I assumed that what I saw was all that was there.

From now on, I fly every pass with the expectation that something is hidden - because sometimes, the most dangerous wires are the ones you've already convinced yourself don't exist.

Trust your gut. Question the obvious. And if the field feels too "clean," look again.

NOTES:

THE WIRE I MISSED
GRUMMAN-SCHWEIZER G-164B

It was already warm when I lifted off that morning, the Arkansas air thick with humidity. The sky was clear, the visibility stretched for miles, and everything about the day felt right. I was flying a Grumman-Schweizer G-164B - a rugged old workhorse I knew well. With 1,900 hours logged, 1,100 of them in this exact aircraft, I'd flown these fields near Sedgwick more times than I could count.

The plan was simple: depart from Bono and circle back to cover a crop field nearby. I'd done it before, and nothing about it felt unusual. The aircraft, though aged, was ready - over 22,000 hours on the frame and still running strong on its Honeywell turboprop engine. I was confident in the machine and comfortable in the mission.

But familiarity has a way of making you blind to the things you think you already know.

At 8:15 a.m., I lifted off and turned toward the first field. Light wind, clear sky, smooth air. I lined up for the first spray pass. Descended low. Focused on the spray line and made the final adjustments. The green expanse stretched ahead - nothing but rows of crops and what looked like open air.

Except it wasn't.

As I began the pass, the aircraft struck a power line strung across the entrance to the field. I hadn't seen it. Or maybe I'd forgotten it was there. Either way, it was too late.

The hit was sudden and violent. The wire grabbed hold of my forward momentum and pulled the airplane downward. The lift disappeared, the stability went with it, and I lost control almost immediately.

We hit the ground hard.

The G-164B slammed into the rows with brutal force. Both wings crumpled. The tail twisted. The fuselage snapped. Debris filled the air. But somehow, there was no fire. No explosion.

I survived.

Bruised and rattled, I crawled from the wreckage.

The air was still, the engine silent. The aircraft lay broken behind me, but I was walking. It felt like a miracle. I didn't need medical help. Just time to process what had just happened.

The investigators arrived shortly after. The aircraft showed no signs of mechanical failure. The engine had been running normally. The controls had been responsive. It wasn't a systems issue. It was a situational one.

The power line was there. I hadn't avoided it. *That* was the cause.

They ruled it what it was: failure to maintain obstacle clearance.

It sounded clinical, but it was true.

In ag flying, the line between success and disaster is razor-thin.

On this flight, I missed that line.

The aircraft was severely damaged. It wouldn't fly again without major repair. But I walked away, and that meant I got to learn from it.

Lessons Learned:

We all know wires are the most dangerous threat in low-level flying.

Unlike towers, wires don't broadcast their presence. They don't flash or reflect.

They blend into the landscape, hide in the horizon, disappear into backgrounds.

And worse, they're familiar. Until they're not...

<u>NOTES:</u>

THE WIRE I SHOULD HAVE SEEN

ROBINSON R66

August in Minnesota brings long days and lots of flying. I'd been in the seat since early morning, spraying fields with the R66, hopping from one job to the next. The rhythm was steady - load, spray, return. The helicopter was performing well, and I was feeling sharp. But the truth is, after so many fields, they start to blur. You trust your muscle memory more than your checklist.

It was during a pass over one of the last fields of the day. I was low - where I needed to be - lined up, focused, everything humming.

Then I saw it.

A power line. Right in my path.

I dropped the collective hard, trying to dip below it. It was an instinctual move, pure reflex. But it wasn't enough. The blades caught the wire, and before I could process what happened, the helicopter slammed into the ground.

I don't remember the impact as much as I remember the silence after. No rotor noise. Just the smell of dirt and fuel. I was lucky - no fire, no injuries. Just the twisted wreckage of a machine that had flown flawlessly up until that moment.

The blades were wrecked.

The vertical stabilizer was torn.

The landing skids were twisted like paperclips.

There were no mechanical failures. Nothing wrong with the aircraft. The only thing missing that day was a complete field recon.

I hadn't scouted thoroughly. I'd assumed I knew the area. I thought I'd remember if there was a hazard.

I was wrong.

Lessons Learned:

In ag flying, every field is a new mission - even the ones you've flown before.

Power lines don't move, but your memory can. One missed recon, one skipped orbit, and you're betting your life on the idea that nothing has changed. But terrain doesn't care about your routine. It only respects preparation.

That day I learned the hard way: a power line doesn't need to be invisible to be missed - it just needs to be ignored.

Now, no matter how many times I've flown a route, I circle the field. I study the lines, the poles, the angles. Because if you're not absolutely certain what's out there, you're not ready to be in it.

What you don't see can destroy an aircraft. And what you don't check can change everything.

NOTES:

THE WIRE I THOUGHT I MISSED
GRUMMAN AG-CAT TURBO

It was a clear morning in Michigan, with 15 miles visibility and calm conditions - an ideal day for agricultural flying. I was flying a Grumman AG-CAT on a crop-dusting mission near Edmore, preparing for a northbound spray pass along the western edge of a potato field. As was common in this area, a set of power lines and telephone poles ran parallel to the field boundary, with an additional wire stretching across the middle of the field from west to east.

I knew the wire was there. I had seen it during my initial recon and planned to pass under it during this spray run. But as I approached, I became uncertain about the exact location of the wire. The aircraft's top wing blocked my view. I thought I had already gone beneath the wire, so I began to pull up at the end of the pass.

Then it appeared - directly in front of me.

There was no time to react. The aircraft struck the wire with both the propeller and upper wing. The wire snapped. Instantly, I felt the engine begin to run rough, and I lost most of my power. I made a split-second decision to perform an emergency landing in another nearby potato field to the northwest.

The field was muddy. As soon as the aircraft touched down, the wheels dug in, and the AG-CAT flipped onto its back. Thankfully, I walked away uninjured. The FAA later categorized the event as an incident.

Lessons Learned:

This was a classic case of losing situational awareness at the worst possible moment. The top wing had obscured my view of the wire during a critical phase of flight. Relying on memory and assumptions rather than visual confirmation was a costly mistake.

In hindsight, the safer approach would have been to fly over the wire in the middle of the field and away from it, rather than attempting to pass underneath. Maintaining a clear visual reference to obstacles - especially those that blend into the environment - is critical.

In agricultural aviation, the margin for error is razor-thin.

Wire strikes remain one of the most dangerous hazards we face.

Plan conservatively, maintain visibility of known obstacles, and never assume you've cleared a threat unless you can see it with your own eyes.

NOTES:

THE WIRE IN THE TURN
GRUMMAN AG-CAT G-164B

It was a calm, sunny morning near Abbeville, Louisiana, and I was flying my Ag-Cat G-164B on an aerial spray run. Powered by a PT6A-34 engine, it was a strong and familiar machine. I had flown a few hundred hours by then - close to half of that in this aircraft - and the mission was routine. Take off from the strip near Kaplan, spray the field, and head back.

I'd already completed several passes - tight, low, and focused. The kind of flying where every second and every turn mattered. I knew the terrain well and had flown this field before.

Coming out of the final pass, I started my turn at the end of the field, ready to enter the pattern again. I wanted it to be tight, efficient. Just a simple swing around to finish the job.

But then, in a flash - everything changed.

A jolt ran through the aircraft. The right wing dropped slightly. I realized I'd hit something.

I hadn't seen it - couldn't see it. A power line had stretched across the far edge of the field. I must've flown right into it mid-turn. The lower right wing and the upper right aileron had made contact.

I quickly leveled out.

The aircraft was damaged but still flying. I assessed quickly: no fuel leaks, engine sounded normal, controls responsive enough. There was no question now - I had to stop. I wasn't finishing the job. I was getting home.

I turned gently, giving the airplane as much stability as I could manage, and headed back to the private strip.

Somehow, I made it back in one piece.

The landing was uneventful, but when I stepped out and looked at the wing, the damage was obvious - substantial, but repairable. I was uninjured, but I knew this was the kind of mistake that stays with you.

Later, in the quiet of the hangar, I thought back to the moment. That wire wasn't new. It wasn't hidden by terrain. It had always been there. I'd even seen it on prior runs. But in the rush of flying, scanning, spraying, and managing the aircraft - my attention slipped. It blended into the background.

The investigation found no mechanical issues. The aircraft performed as it should. The engine was strong. There was no gusty wind, no distracting radio call - just a moment of visual miss. And in ag flying, a moment is all it takes.

Lessons Learned:

This kind of accident happens more often than it should, and for the same reasons. Wires are difficult to spot, especially in certain light. And the more familiar the field, the easier it is to let your guard down. The workload at low level is intense - systems, terrain, timing, focus - it's all happening fast.

Here's what I learned, and what I'll never forget:

- Re-brief every field before every flight, even if it's your tenth pass. Especially near the turn points.
- Use GPS overlays or field maps to mark hazards - don't rely on memory.

- If there's any uncertainty, pull up and survey before you commit to the next pass.
- Talk to ground crew or locals about any changes - new wires, new poles, new risks.

After the strike, I kept my cool and got the aircraft home safely. But the strike itself shouldn't have happened. And the truth is, it could've been worse.

In ag aviation, it's not always the hard flying that gets you. It's the quiet traps - the wires that wait, invisible in the light, until one turn cuts just a little too close.

The field didn't change. I just didn't see what was already there. And that's a mistake I'll only make once.

NOTES:

THE WIRES I DIDN'T SEE
AIR TRACTOR AT-802

It was mid-July in Minnesota, late afternoon, and the light had taken on that familiar golden hue. I was flying my AT-802, ready to knock out an application job just a short hop from Benson. The air was warm, a bit gusty, but manageable. What made things tricky was the smoke - thin, hazy layers carried in from distant wildfires. Visibility wasn't terrible, but it wasn't great either. Everything looked a little muted.

As I approached the field, I circled twice to get my bearings. I scanned for obstacles, especially wires. I spotted two power poles on opposite ends of the field, spaced fairly wide. I took them as the ends of a span. That was my mistake. I assumed there was no line strung between them - or perhaps I just convinced myself there wasn't.

With the field scoped and the wind aligned, I set up my pass, low and tight. I settled into the rhythm that thousands of hours had etched into muscle memory. The AT-802 hummed below me, heavy and steady.

Then, *snap.*

The aircraft jolted violently. I'd hit the wire. A high-tension line strung invisibly across the field, right between those poles.

The right aileron took the blow. I felt the aircraft yaw and shudder under the sudden load. But it held together.

I broke off the run, climbed out, and nursed the airplane back to Benson. The landing was smooth, all things considered. Once on the ground, I got out and inspected the damage. The right aileron was wrecked. The rest of the aircraft was bruised but intact. I was lucky.

No mechanical issues. No warning lights. Just a simple case of misreading the scene.

I'd done what I thought was a thorough field survey - but the haze had dulled my visibility, and my own assumptions had filled in the rest. Those two poles weren't the end markers I imagined. They were the anchors of an active line - and I flew right through it.

Lessons Learned:

In ag flying, our greatest risks don't always come from malfunction - they come from misjudgment. And often, the margin between safe and sorry is a single overlooked detail.

Wires are invisible killers, especially under smoke, haze, or bright light. Even seasoned eyes can miss them. And when we convince ourselves a hazard isn't there, we stop looking for it.

That day reminded me that a field check isn't complete until every possibility is eliminated, not just assumed away. If there's a pole, there's a reason it's there. If there's haze, it's hiding something.

I walked away from that flight. The airplane didn't leave me stranded. But it gave me a lesson I won't forget.

Always fly as if the wire is there - because sometimes, it is.

NOTES:

THE WIRES I KNEW WERE THERE
AÉROSPATIALE SA 315B LAMA

By the time I lifted off that morning, the sky over Idaho was crystal clear. I'd reviewed the hazards. I'd studied the field. I'd flown around it before. This wasn't unfamiliar territory - it was a straightforward aerial spray job on terrain I knew well.

I was flying an SA315B, a rugged workhorse of a helicopter. I'd logged nearly 2,000 hours in my career, with over 500 in this aircraft alone. At 32, I wasn't new to the game. I'd departed Cottonwood around 9:30 a.m., headed toward the fields near Lewiston. Visibility was unlimited. The air was still. It was a perfect morning for ag flying.

I'd already walked through the plan in my head. I knew the location of the power lines that ran across the job site. I'd seen them with my own eyes. Marked them on my tablet. I'd made several passes around them that morning without issue.

In ag flying, every field has its rhythm. You get into the groove - pass after pass, line after line, always keeping one eye on the terrain and the other on the wires, trees, fences, and wind. It becomes instinctive. And that instinct is both what keeps you alive and what can get you hurt.

I got too comfortable.

I can't say exactly what happened. Maybe I turned just a little early. Maybe the light shifted and blended the wire into the horizon. Maybe I trusted my mental map a little too much. But on one of those passes - just one - I missed.

The rotor struck the power line.

The impact was immediate. There was no question what had happened. The aircraft shuddered. The main rotor caught the wire and threw the whole system into chaos. The transmission took the brunt of the force. The tail rotor drive shaft snapped. The tail boom buckled. Control was slipping fast.

There was no time to think. I didn't check instruments. I didn't troubleshoot. I just reacted. I dropped collective and committed to a forced landing right there in the field.

We hit hard.

The skids slammed into the ground. The main rotor shattered. The tail boom twisted. The transmission was toast. But somehow, I was okay. Bruised, sure. Shaken, no question. But alive. And I walked away.

There was no fire. No explosion. Just silence.

When the investigators came, I told them the truth. The aircraft had been running fine. No engine problems. No control issues. No warnings. The only thing wrong that day was me. I hit a wire I already knew was there.

They confirmed it. The rotor blades were splintered. The transmission casing was fractured from the impact. The tail rotor and drive shaft were twisted, mangled.

The cause of the crash? Pilot failure to maintain clearance from a known obstacle.

That's it.

Plain and painful.

Lessons Learned:

Wires are the oldest enemy in low-level flying. Everyone who flies ag knows this. We've all seen what happens when someone hits one. And we all carry that fear tucked somewhere behind our confidence.

But here's the truth: the most dangerous wire is the one you already know is there. The one you've flown around all morning. The one that fades from warning to background.

The lesson I learned the hard way is this:

- Don't ever relax around a wire.
- Don't let familiarity blur caution.
- Don't rely on memory alone.
- Reinforce known hazards with visible markers if you can.
- Fly every pass like it's the first.
- Stay deliberately separated - laterally and vertically. Always.
- Because the moment you think "I've got this," is the moment the wire finds you.

I was lucky. My seatbelt held. My reflexes kicked in. The ground was forgiving enough. But that doesn't change what happened. I got complacent. And the helicopter paid for it.

So here's my advice - don't just remember where the wires are. Respect them every time you fly past. Because they're not waiting for you to forget. They're just waiting for you to look the wrong way for one second.

And when that happens, they don't give second chances. Only lessons.

NOTES:

THREADING THE WIRES
AIR TRACTOR AT-602 F

The sky was clear, the fields stretched wide, and the afternoon light made everything seem straightforward. I was flying an Air Tractor AT-602 for a spray job outside New Washington, Ohio. I'd already reviewed the layout from above, flown a recon pass, and planned my pattern - north to south, right under a set of high-tension power lines that crossed the field.

I held an airline transport certificate and had logged nearly 10,000 hours. The aircraft had recently passed its 100-hour inspection. Everything felt solid. Conditions were great. The plan was efficient: spray beneath the wires, reduce turning, keep things clean and simple.

The first pass went well - just feet above the canopy, right beneath the wires. No issues. Smooth and steady.

Then came the second pass.

As I lined up, I could feel how tight the margin was. Flying low is always unforgiving, but the light had shifted slightly, and visibility of the wires wasn't as clear as before. I continued the pass, watching my height, but somewhere in the transition, I misjudged.

Midway through that pass, the rudder hit the wire.

The impact was immediate. The aircraft jolted, yawed sharply, and I lost all directional control. The rudder was gone - severed - and the aircraft was suddenly unstable, with no way to correct its heading.

I tried to hold it together, aimed for the field, and managed to bring it down. Not gracefully. We hit hard, skidded sideways, and finally came to rest - damaged, but upright. I climbed out with minor injuries, shaken, but alive.

There'd been no mechanical issues. No warnings. No strange behavior before the impact. The engine was strong. The systems were fine. It was a simple mistake - too low, too close, and not enough room to recover.

In the days that followed, I replayed it over and over. The lighting, the angle, the way wires disappear at certain times of day. It doesn't take much. Inches make the difference.

The aircraft had no ELT onboard, but that wasn't the issue. The weather was perfect, the field familiar. What brought the aircraft down was a failure to maintain that critical clearance. The difference between a routine pass and a wreck was a moment and a margin.

Lessons Learned:

Wires are still the most dangerous part of ag flying. And they're everywhere - silent, unforgiving, and often invisible until it's too late.

What I learned that day reinforces what every low-level pilot should know:

- Wires don't always show up, even on a clear day. Look for poles, not cables.
- Recon passes aren't just procedural - they're survival tools. Use them thoroughly.
- Plan your spray pattern to minimize wire crossings. Efficiency means nothing if it puts you at greater risk.

- Never assume one clean pass guarantees the next one will be the same.
- Fatigue, changing light, and familiarity can erode focus in subtle but deadly ways.

I'd flown thousands of hours, across hundreds of fields. I'd seen more wires than I could count. But that day, I saw how easy it is to misjudge - how fast things go wrong when the margin is gone.

In this job, the difference between success and disaster is measured in feet. Sometimes inches. The danger isn't the field. It's the wire you thought you cleared.

Respect every one of them.

Know the route. Scan every pass. And never forget: in ag flying, precision isn't just about spray patterns. It's about staying alive.

NOTES:

TOO CLOSE TO THE BRANCHES
BELL 206B JETRANGER

It was a hot summer evening, and I was wrapping up a full day of aerial spray runs in my Bell 206B JetRanger. The helicopter had been performing flawlessly. The systems were solid, the weather was calm, and daylight still hung in the air as I approached the fuel truck for a quick top-up.

I'd done this landing dozens of times before - onto the trailer near our staging area. The spotter confirmed the area was clear. Routine stuff. At least, that's what I thought.

As I made my approach, descending steadily, I focused on the landing zone. Everything looked fine - until it wasn't.

I felt it before I heard it. The tail rotor clipped the limbs of a nearby tree. Just like that, multiple blades separated from the hub.

The aircraft lurched. I lost tail rotor authority almost instantly. The helicopter began yawing hard - swinging under the torque of the main rotor. I knew I couldn't land on the trailer anymore. I aimed for the open ground beside it instead.

But before I could get there, the entire tail rotor system - gearbox included - ripped away from the airframe.

Now I was in a spin.

There was no more directional control. I fought to keep it steady, to cushion the descent as best I could. Somehow, I managed to bring it down hard, but upright. The helicopter skidded and came to a stop.

It was over in seconds.

I stepped out unhurt. The aircraft was badly damaged, but there was no fire, no secondary crash. Just bent metal and a lot of relief.

In the aftermath, the inspection was clear. No mechanical issues, no pre-existing failures, no red flags in the logs. The 100-hour inspection had been done just over a week before. Everything had worked - until I got too close to those branches.

The initial contact with the tree limbs had set off the chain reaction. The blades struck, then separated. The gearbox followed. And the rest was just physics.

The spotter had said it was clear. And I believed him. But "clear" needs to be more than a checkbox. Because in that moment, I'd underestimated just how much space the tail needed - and how easy it is to lose track of what's behind you when all your focus is forward.

Lessons Learned:

Rotorcraft operations around trees and trailers demand more than just hours in the logbook. They demand vertical awareness - especially behind the cockpit.

Here's what this flight taught me:

- Tail rotor clearance is critical. What you can't see can still bring you down.
- Vertical hazards - like overhanging limbs - are just as dangerous as lateral ones. Don't assume they're out of the way unless you've confirmed it.
- Spotters are essential, but their perspective isn't always perfect. Trust them - but verify when you can.

- Routine can dull your senses. Familiarity with a landing zone doesn't make it safe. It just makes the risks easier to forget.

I walked away. The aircraft didn't. And I'm still grateful I had the training and presence of mind to make the emergency landing work.

But the truth is, I never should've been that close to the branches.

In helicopters, the smallest contact can turn into a major emergency. You may not feel it right away. But by the time you do - it's usually too late to fix.

The margin for error in rotary-wing flying is razor-thin. And when it comes to the tail rotor, that margin almost doesn't exist.

Know your space. Respect the invisible. And never assume that just because you've landed there before, you can afford to get complacent.

NOTES:

TOO CLOSE TO THE LINE
THRUSH S2R-H80

The field looked straightforward. I'd flown it before - flat, clear, with only one major hazard: a power line running nearby. As always, before spraying, I did my two surveillance passes. It was a habit, one I believed would give me the edge in safety. The first pass went smoothly. On the second, I dipped in closer to get a better angle for the job ahead.

I knew the wire was there. I'd seen it on the first pass. But somehow, on that second run, I lost track of it. Maybe the light shifted, or the background camouflaged the cable - whatever it was, I didn't see it again until it was too late.

There was a jolt, then a wrenching lurch as the landing gear caught the line. Instantly, the aircraft nosed over and slammed into the ground. The wings tore apart on impact. The fuselage buckled. My world flipped from forward flight to twisted metal in seconds.

I was lucky. I climbed out with only minor injuries. The aircraft, though, was in ruins.

In the aftermath, the story was simple: there were no mechanical issues. The engine had been running fine, the aircraft airworthy. The weather was clear. The problem wasn't the plane. It was the pilot.

I hadn't ensured clearance from the obstacle I already knew existed. I'd flown that pass believing I could thread the gap safely, relying too much on familiarity and confidence. The cost was heavy.

Lessons Learned:

Knowing a hazard is there is not the same as actively avoiding it.

Wires are a killer in ag flying. They blend into backgrounds, hide in plain sight, and punish hesitation. I'd told myself I was being cautious with my two-pass system. But I failed to treat that second pass with the same precision as the first. I assumed familiarity equaled safety. It didn't.

Now, I make it a point to mark wires mentally - and visually - every single time. I talk them out loud, confirm with GPS overlays, and never let routine dull my edge.

Because in this business, one forgotten wire can end it all.

NOTES:

UNSEEN, UNFORGIVING
UNSPECIFIED AGRICULTURAL TURBOPROP

The morning was typical for aerial spraying - clear skies, 5-mile visibility, and a routine mission near the edge of Ririe, Idaho. I had flown hundreds of fields like this before. My aircraft was loaded and ready for a pass over a potato field tucked near town. It wasn't my first time spraying in this area, and I conducted a standard airborne recon before beginning the run. Everything looked clear.

Or so I thought.

As I made my first pass, low and fast over the crop, I struck something - hard. A jolt, a crack, and a shattered windscreen told me the story. I had hit a power line that stretched across the center of the field - one that I never saw, not even during the aerial inspection.

I later learned that the line had long since been decommissioned and wasn't supplying electricity to anything. It was still suspended, though, connected to a pole hidden in the dense foliage of a tree dead-center in the field. From the air, the tree and the wire had blended perfectly into the background, and with no visible poles or obvious anchors, the hazard was virtually invisible.

Thankfully, the damage to the aircraft was limited to the windscreen, and I walked away uninjured. But the incident could have been far worse.

Lessons Learned:

This close call was a powerful reminder that in ag flying, the threats you can't see are the most dangerous.

Despite a pre-run aerial recon, I missed a dormant power line hidden in foliage. Proof that even familiar fields can hide deadly surprises.

From now on, I won't rely solely on what looks clear from the cockpit. If there's any uncertainty - especially near town edges - I'll walk the field, question the grower directly, and treat every tree or pole as a potential anchor for the unseen.

I've also learned to advocate for marking dormant infrastructure. Because once you've hit an invisible wire, you realize that seeing it before the impact is the only safe option.

NOTES:

VEILED DANGER

CESSNA 188 (AG WAGON)

It was an overcast Saturday morning - one of those calm, hazy starts where the high layer of cirrus clouds dims the sun just enough to blur the shadows. I was out early in my Cessna 188 Ag Wagon, preparing to lay down a smoke pass to confirm the wind direction before beginning a crop application.

I had flown this field countless times before. I knew where the power lines were - or so I thought. My plan was simple: descend over the wires, level out just above the deck, and drop smoke. It's something I've done thousands of times. On this run, I was actually a little higher than usual on approach - just to be cautious.

But with the diffused light from the overcast sky, my depth perception was off. I couldn't see the wires clearly against the gray backdrop. I believed I was high enough, trusting my experience and spatial judgment. But the wires appeared - just as I struck them.

The aircraft jolted, and I immediately pulled up to assess what had happened. I knew I'd made contact - my prop and the wire cutter mounted on the windshield had done their job. I gained altitude and checked for any irregularities - no vibration, no strange noises, no warning signs.

I circled back and confirmed the power line was down. Visually scanning the aircraft again, I didn't see any damage. Feeling it was still safe, I completed the application before returning to the airstrip for a full inspection. A few minor scratches were all I could find. I reported the strike to the power company immediately.

The culprit? A combination of poor wire visibility due to the light conditions and an unusually high wire strung across a drainage draw - higher than the typical lines I'd grown used to clearing.

Lessons Learned:

Even in well-known fields, light conditions can change everything.

That overcast sky didn't just dim the day - it blurred the wires into the haze and dulled my depth perception.

I trusted memory and instinct, but both failed under veiled visibility.

It was a humbling reminder that no amount of experience can replace fresh caution.

From now on, I'll give every field a fresh look - especially under changing light.

A wire strike may not always bring disaster, but it always brings risk. Margins are measured in feet, and feet vanish fast when visibility fades.

Know the conditions. Know the layout. And when in doubt, circle once more.

It's worth the extra pass.

NOTES:

WHAT I DON'T SEE CAN HURT
UNSPECIFIED AGRICULTURAL LOW-WING AIRCRAFT

It was the second field of the evening - a routine spray job in Nebraska, or so I thought. With over 2,300 hours in my logbook and six years of ag flying behind me, I followed standard procedure: circle the field, locate obstacles, mark the power lines. Two sets were visible - one running east-west along a road, the other running north-south. I noted the convergence point in the southeast corner of the field and set up my approach.

What I didn't see was the third set: four #2 gauge wires strung diagonally from northeast to southwest - right through that same corner. No poles. No warning. Just wires suspended invisibly between the two poles I had already marked.

As I began my pull-up to clear the east-west lines, the aircraft jolted. I'd hit the hidden wires. There was no time to react. The prop chewed through three of them; the fourth struck the cockpit, slashing into the vertical and horizontal stabilizers. One wire hit the safety cable connecting the cockpit to the tail. Somehow, the aircraft kept flying.

I didn't wait around.

I made a beeline back to Grant Airport.

On inspection, the damage was clear: three nicks in the prop, a puncture in the left stabilizer, skin damage to the vertical stab, and a shredded safety wire. The aircraft repairs would total about $2,500. The utility company estimated their damage at $667.

But the true cost was a wake-up call. I'd done what most ag pilots do - looked for poles, not wires. I expected the field to match the visible setup. But I missed the hidden hazard - the one that didn't fit the pattern.

Lessons Learned:

I did everything right - circled the field, marked visible wires, spotted the poles. But I missed one thing: four nearly invisible wires strung diagonally across the southeast corner.

On my final pull-up, I struck them. The prop sliced through three; one ripped into the tail. Somehow, I flew home.

Damage was fixable. But the lesson was lasting.

Not all threats come with poles or patterns.

Now I ground-scout every field and ask landowners directly. In low-level ag flying, assumptions are dangerous. What you don't see can hurt you.

And sometimes, it's the wire you never expected that leaves the deepest scar.

NOTES:

WHEN THE WIRE WINS
UNSPECIFIED AGRICULTURAL AIRCRAFT

It was a hot, breezy evening near our Airport in Delaware. I was finishing a spray job in a field that, like many I've flown over in my agricultural career, had a power line stretched across it. The layout of the field made one thing clear: the most effective way to spray required flying underneath the wire.

Operating beneath wires isn't unusual in this line of work, but this one was lower than most, leaving almost no margin for error. I had made several passes successfully, flying low and tight, adjusting for the swirling wind and the heat-blurred horizon. But on the final pass, something changed.

As I flew beneath the wire for the last time, the aircraft suddenly ballooned upwards. Whether from a thermal, a gust, or just a moment of lift catching me off guard - I felt the bump. The vertical stabilizer had clipped the wire.

Miraculously, the damage appeared minimal. I was still in control, and I made the call to return to the airfield. I landed without further incident and inspected the aircraft. The stabilizer had indeed made contact, but the damage was light. Still, it could've been much worse.

Looking back, I realize the decision to fly under that wire in marginal conditions wasn't wise. The combination of heat, wind, and the low height of the wire made the risk unacceptable. The pressure to get the job done, especially when the farmer is waiting, often pushes us beyond what's safe. But the field could've waited, or we could have approached it differently.

Lessons Learned:

Familiarity breeds confidence, but also risk.

Flying under wires may be part of ag flying, but this incident reminded me that "routine" doesn't mean "safe," especially when conditions shift.

Heat, gusty wind, and swirling thermals all conspired in that moment, lifting me into a wire I'd already passed safely multiple times. The margin was razor thin, and I pushed it too far.

From now on, I'll factor in environmental changes more critically and reassess each pass based on current conditions, not past success.

No spray job is worth risking the aircraft or my life.

If the wire is too low or the air too unstable, the job *can* and *should* wait.

<u>NOTES:</u>

WHIRLWIND AT THE WIRE
UNSPECIFIED AGRICULTURAL AIRCRAFT

It was supposed to be the final pass of the day . One more run across the end of a field near 9NE1 in Nebraska. I'd been flying all afternoon, completing spraying operations in mostly good visibility. The only factor that stood out was the sporadic appearance of dust devils - whirlwinds that would swirl up from the earth and vanish just as fast. I'd seen a few that day, but this next one was hiding.

As I pulled up at the end of the pass, I scanned visually and cleared the power lines ahead. That's when everything changed.

Without warning, the aircraft suddenly yawed hard left and dropped its nose. I slammed full right rudder, trying to fight it back under control. But by then, I was already in the wires. There was no time to roll out or gain altitude. I did what I could - straightened the plane just before impact.

I heard and felt the hit. The right wing leading edge took a glancing blow from the wire, about 18 inches long. The nose bowl had some paint scraped off, and I broke two wires. I radioed the local power company within 11 minutes of the strike.

This invisible whirlwind had formed in an irrigated cornfield. No dry debris, no dust plume to alert me.

I didn't see it until it was too late. Most whirlwinds are visible if they're kicking up dirt, but when they form over wet ground? You won't know they're there until they hit you.

Fortunately, the aircraft was still airworthy. I completed the flight and landed safely.

Lessons Learned:

I cleared the wires, but didn't see the whirlwind.

Hidden in an irrigated cornfield, it had no dust, no warning. In an instant, it yanked my aircraft left and down. I fought for control, but the wing struck the line.

The damage was light, but it shook me.

When you're flying low and heavy, even a small gust can tip the balance. Whirlwinds over wet ground don't announce themselves. If you spot a few, more are likely. Don't gamble on the last pass.

The riskiest threats are often the ones you can't see.

NOTES:

PART 2 - HUMAN FACTORS & DECISION MAKING

"Airspeed, altitude or brains: Two are always needed to successfully complete the flight."
–Anonymous

HAZE CLOUDED JUDGMENT
UNSPECIFIED AGRICULTURAL AIRCRAFT

I was in the cockpit of a small agricultural aircraft, engaged in aerial application. The day was challenging: haze from nearby wildfires had reduced visibility to just 4 miles, and the airspace was turbulent. As I approached a highway at low altitude, I knew I was about to encounter a significant obstacle: a cross-country power line. The aircraft's weight and the turbulence meant climbing to clear the wires wasn't an option, so I had no choice but to plan a maneuver that would take me under the power line.

As I neared the highway and the wires, my initial plan was to climb, but I quickly realized that I wouldn't have enough altitude to clear the power line. That's when I made the decision to fly under the wires. The challenge? The highway traffic, which I hadn't anticipated. Flying at low altitude so close to traffic was uncomfortable, but it was my best option to avoid hitting the wires.

At that moment, I started a left turn to avoid the traffic below and position myself better to pass safely under the wires. The decision wasn't easy, and I was surprised at how little space there seemed to be between the power lines and the road. What contributed to this misjudgment?

The haze.

With the smoke from the wildfire blending the power lines into the haze, it was difficult to gauge their exact location. I had initially believed they were farther away than they actually were.

After that close call, I made an immediate adjustment to my approach. I began giving more space to power lines, pulling up earlier and adjusting my maneuvers to ensure greater safety. This event underscored the importance of situational awareness - particularly when weather conditions like haze and smoke distort our perception.

Lessons Learned:

Haze doesn't just blur the horizon - it clouds judgment and distorts critical depth perception. That day, wildfire smoke turned familiar terrain into a deceptive landscape. Power lines I normally would've anticipated and cleared easily appeared farther away, blending almost invisibly into the muted backdrop.

When visibility is reduced - even to just four miles - our instincts and training are put to the test. But the truth is, no amount of experience can override physics or trick perception into seeing clearly through smoke. I misjudged distance, had to adapt quickly, and ended up maneuvering beneath wires uncomfortably close to highway traffic.

Since then, my rule has changed: in any kind of haze, I plan for more altitude, more time, and more room than I think I need. I don't assume I'll see the wires in time - I assume I won't.

When the air is thick with smoke, the safest approach is margin over muscle. Give obstacles more space, because in poor visibility, what you think you see can be fatally wrong.

NOTES:

FINAL MOMENTS, FAST DECISIONS
AIR TRACTOR AT-402

After performing a wheel landing on the runway, I used the partial beta setting on the torque lever to help with deceleration. I had previous experience flying an AT-402 with a PT-6 engine, where I frequently used beta for this purpose. However, I had recently switched to an AT-402 with a Walter engine, and I had been told that the beta setting on this aircraft was more difficult to engage - it was stiffer than what I was used to.

As I tried to adjust the torque lever, I applied more force than anticipated to push it past the detent. The lever unexpectedly went to full beta, which caused the aircraft to veer left. I immediately tried to correct the situation by applying right brake while also pushing the torque lever back out of beta. I managed to disengage the beta setting and prevent a ground loop, but by that point, the aircraft had already deviated about 30 degrees from the centerline.

At this point, I applied both brakes and brought the aircraft to a stop just off the runway surface. Unfortunately, the aircraft's spray boom, which hangs approximately one foot from the ground, struck and damaged a taxiway sign. I immediately reported the damage to the FBO and the Airfield Point of Contact.

Lessons Learned:

Transitioning between similar aircraft can be deceptive. While an AT-402 may look and fly like its PT-6-powered counterpart, subtle differences - in this case, the Walter engine's torque lever - can lead to major surprises.

My mistake wasn't simply using beta - it was assuming I could use it the same way I always had. The stiffness of the Walter engine's torque lever caught me off guard, and when I applied extra force to engage beta, the lever jumped past where I intended. The result was a sharp, unexpected yaw and a brief but intense loss of control on rollout.

Familiarity breeds comfort, but it can also breed complacency. I should have practiced smaller, gradual adjustments on earlier landings or avoided beta altogether until I fully understood its behavior in this specific aircraft.

The lesson is clear: even small mechanical differences matter. Take time to relearn the aircraft - especially the systems that impact ground handling. Because when it comes to rollout control, there's very little room for overcorrection.

NOTES:

THE CHECKLIST MISS
AIR TRACTOR AT-602

I'm submitting this report because I believe it highlights a critical "human factors" event that led to a mishap. I had not flown any ag aircraft for about 90 days, having spent the previous two weeks instructing a multi-engine student in his own airplane. There were several distractions leading up to the mishap, each contributing to a breakdown in discipline, resulting in a critical checklist deviation and landing without the tailwheel lock engaged.

First, I had not done a refresher on operations procedures before picking up the airplane after its annual inspection. While I had over 500 hours in this aircraft in 2019, I had been flying Air Tractor aircraft with a different tailwheel system, where the tailwheel locks automatically when the stick is aft of a certain position. This habit likely led to my omission of the tailwheel lock check in my normal pre-landing flow. Additionally, my focus on instructing in a dissimilar aircraft over the last couple of weeks may have caused a lapse in my typical discipline.

During the 15-minute flight home, the right cockpit door latch vibrated loose, distracting me. I was concerned about it opening mid-flight, which could have resulted in its separation from the airplane.

This distraction further diverted my attention from my usual pre-landing safety checks.

Another factor was the recent maintenance on the aircraft, which included new brake discs and pads. These components typically require 2–3 landings to break in, and their reduced effectiveness was on my mind as I approached the landing. Concerned about the softness of the brakes, I mentally prepared for it, which again distracted me from completing my routine checklist.

As a result, I forgot to check the tailwheel lock before landing, and once I selected beta thrust to decelerate during the landing roll, the free-swiveling tailwheel severely compromised my ability to control the aircraft. This resulted in a runway excursion and a ground-loop.

In hindsight, I noticed there was no accessible checklist in the cockpit of the aircraft. I intend to rectify this by placing the relevant checklists from the Aircraft Flight Manual in full view in the cockpit for routine use. This experience serves as a reminder that even seasoned pilots, when distracted or out of practice, can make critical errors that can have serious consequences. I now understand the importance of adhering to checklist discipline, not just for safety, but to set an example for others.

Lessons Learned:

No matter how many hours we log, discipline - not just experience - is what keeps us safe. Returning to the AT-602 after a break and recent time instructing in a multi-engine aircraft, I underestimated how easily habits can blur between platforms. My oversight? Failing to lock the tailwheel prior to landing.

Distractions stacked up: a loose cockpit door latch, concern over fresh brakes, and assumptions formed from flying other models where the tailwheel locked automatically. Each seemed minor in isolation - but together, they eroded my checklist discipline. The result was a ground loop that could've been avoided by one small action.

What struck me afterward was the absence of a visible checklist in the cockpit - something I'm now correcting immediately. Routines protect us, especially when we're rusty, rushed, or distracted.

The takeaway is this: experience is not immunity. When your flow gets interrupted or your platform changes, lean harder on checklists, not instincts. Safety lies not in what you remember, but in what you verify - every single time.

NOTES:

OVERCAST AND OVERCONFIDENT
UNSPECIFIED AGRICULTURAL LOW WING AIRCRAFT

During a spraying operation, I struck power lines due to overcast conditions that made the wires difficult to see, blending into the terrain. The low visibility and minimal contrast between the power lines and the surrounding environment left me with insufficient time to avoid the wires when I first saw them.

Looking back, I recognize that I could have exercised better situational awareness. In overcast conditions, the wires become even more challenging to spot, and I should have taken more time to assess the terrain and look for obstacles before beginning the spraying operation. In this case, better preparation and awareness could have prevented the minor incident, which thankfully resulted in no injuries or damage.

From this experience, I've learned the importance of adjusting my approach when flying in overcast weather. It's vital to be extra cautious, giving myself more time to observe the landscape and identify potential hazards, such as power lines. This lesson is a reminder to prioritize situational awareness in all conditions, particularly when visibility is compromised.

Lessons Learned:

Overcast skies can quietly erase the contrast we rely on to spot hazards. During a spray run, I missed seeing power lines until it was too late - blending into the terrain under dull light. The mistake wasn't just visibility - it was failing to adjust my awareness for the conditions.

In agricultural flying, familiarity isn't enough when visibility drops. Slowing down, reassessing the terrain, and planning for low-contrast obstacles is essential.

The takeaway: overcast doesn't just dim the sky - it hides the threats. In compromised conditions, take extra time. What you don't see can hurt you.

NOTES:

HUMAN ERROR, EMPTY TANKS

AIR TRACTOR AT502B

During a routine agricultural mission, I experienced an off-airport landing due to fuel exhaustion. The day began with standard operations: I was preparing to fly a field three miles east of the airport. The mission required a large hopper load and half a tank of fuel.

After the ground crew loaded the aircraft, I took a brief lunch break, during which I received a letter with negative feedback about my performance. Returning to the aircraft, I found the ground personnel had finished their tasks, and I reset the fuel totalizer to prepare for the flight.

The issue began to unfold about 45 minutes into the flight while I was performing an aerial application. The engine quit mid-turn due to fuel exhaustion, and I was forced to make an off-field landing.

The aircraft's fuel gauges are of the single sending unit type, which can be erratic due to fuel sloshing in the hopper during steep banking turns. Additionally, the low fuel warning light is poorly positioned and difficult to see in bright daylight, especially with the intensity of sunlight interfering with visibility.

Unfortunately, I had placed too much trust in the ground personnel and became distracted by concerns unrelated to the flight, leading to my complacency and fatigue, which were ultimately key factors in the fuel exhaustion.

This experience has prompted the realization that I must take a more proactive role in overseeing the fueling process to ensure it aligns with expected standards. Going forward, I will take steps to verify that proper fueling procedures are followed, reducing reliance on ground personnel to ensure accuracy. As an additional precaution, we are discussing aircraft modifications to improve safety, such as adding a master caution light that will be placed in the pilot's line of sight and equipped with adequate intensity to make warning lights more visible. We also plan to install a refueling system that allows for a partial fuel setting, automatically stopping the flow of fuel once the appropriate quantity has been loaded.

Lessons Learned:

This incident underscores the critical importance of fuel management and human factors in aviation. Fatigue, complacency, and a breakdown in communication between flight crews and ground personnel contributed to the fuel exhaustion. As pilots, we must take an active role in monitoring all aspects of flight preparation, including fueling, and remain vigilant for any potential distractions or external pressures that may impact decision-making.

In future operations, I will strive to maintain greater situational awareness, particularly when managing fuel quantities and the behavior of the aircraft's warning systems, to ensure such an incident does not happen again.

<u>NOTES:</u>

MIXED MESSAGES, EMPTY TANKS

AIR TRACTOR AT-502

While returning from an agricultural spraying mission, I experienced an engine flame-out due to fuel exhaustion. The engine failed to restart, despite attempts to use the igniters and boost pump, so I prepared for a forced landing. Unfortunately, I couldn't glide to a suitable road and ended up landing in a cornfield. Thankfully, I was unharmed, and the aircraft sustained no damage.

Upon investigation, it became clear that the cause of the flame-out was fuel exhaustion. As part of the aircraft loading process, the aircraft is fueled before each mission. During the busy season, the ground crew handles tasks such as loading chemicals, cleaning the windscreen, and fueling the aircraft using a bottom-load system. Sometimes, when the workload is heavy or the crew is understaffed, I assist the ground crew in these tasks. On this particular day, the loader was working alone, and I frequently helped out.

Before the engine flame-out, I had stepped away from the aircraft to make a call to a farmer regarding an irrigation pivot. When I returned, both the chemical load hose and the fuel hose had been disconnected.

We discussed taking a lunch break after the load, and I gave the loader the thumbs-up, asking if we were, "Good to go." He replied, "Good to go!" However, I wasn't actually ,"Good to go," as the aircraft had not been refueled.

The loader later explained that he thought I had refueled the aircraft, as I had assisted with fueling earlier in the day. Meanwhile, I assumed he had refueled it after disconnecting the hose.

A contributing factor to the confusion was that the aircraft I was flying was different from the one I usually fly, which is equipped with an MVP flat panel instrument display. This system provides a digital and tape format of the fuel quantity and an audible warning when the fuel level is low. The aircraft I was flying during the incident only had traditional round-dial gauges, without the audible warning I was used to.

To prevent a recurrence, I will now personally verify the fuel level by removing the fuel cap and visually checking the tank. If I remain in the cockpit, I will ensure the loader or an assistant "sticks the tank" and shows me the fuel stick.

Additionally, we have updated our loading procedures to ensure that whoever disconnects and stows the fuel hose must also verify that the aircraft has been fueled and that the fuel caps are secured before disconnecting the hose.

I will also modify my checklist procedure, moving the takeoff checklist to when the aircraft is stationary, rather than during taxi.

Lessons Learned:

This incident highlights the importance of clear communication between flight crew and ground personnel, especially during busy operational periods. Misunderstandings about fueling procedures, combined with reliance on different instrumentation, led to a critical failure. By implementing more robust procedures and ensuring that both the pilot and ground crew are on the same page, such incidents can be avoided in the future.

In summary, I will ensure I verify fuel levels personally and enforce clearer communication during the fueling process to avoid placing blind trust in assumptions.

Moreover, modifying pre-flight checklists and ensuring thorough visual checks will help prevent similar failures.

NOTES:

SNAP OUT OF IT!
BEECHCRAFT TWIN BEECH 18

Flying an agricultural survey mission over the busy Class C airspace near Riverside, California, our two-pilot crew in the Twin Beech was operating under VFR. I was in the right seat as First Officer and not flying the aircraft at the time. We had just launched on our second flight of the day, working a pattern of north-south passes near the PDZ VOR, a few miles from Ontario Airport.

While inbound on a northbound leg, SoCal Approach came on frequency with a traffic alert: "Traffic, 11 o'clock, moving to 12, same altitude, converging." It was a Skyhawk (C172), and it was coming in fast.

As SoCal gave the call, I spotted the aircraft immediately and replied, "In sight, we'll descend and pass behind." But to my surprise, the captain, sitting left seat and in command, made no move to avoid.

I asked, "Do you see him?"

"Yes," he replied.

"We need to descend."

"Yes," he repeated.

Still, nothing.

The aircraft was closing rapidly.

With the Skyhawk now far too close for comfort, I shouted, "Captain, descend!" and pushed the control column forward myself. We dipped below and passed behind the traffic with what I'd estimate was a barely safe margin. No collision - thankfully - but it was far closer than I'd ever want to repeat.

After the pass, the captain seemed to snap out of a fog. "I'm sorry," he muttered. I didn't say anything - I was too rattled and, frankly, too irritated.

Looking back, I don't believe he meant to ignore the threat. He's a competent captain, but that day something was clearly weighing on his mind. And in the world of aviation, even momentary distraction can put lives in danger.

Lessons Learned:

Even in a two-pilot cockpit, complacency and distraction can quietly creep in.

That day, what saved us wasn't the warning from ATC. It was the insistence to act, to intervene when something felt off. A distracted captain, no matter how skilled, can become a risk if unchecked.

The key lesson? Always stay engaged, and never assume the other person will act. Crew environments demand shared vigilance. If something doesn't feel right, speak up - and if necessary, take the controls. It's not about rank; it's about responsibility.

In the sky, a split second of clarity can be the difference between a safe pass and a headline.

NOTES:

FAMILIAR FIELD, NEW FACES
AGRICULTURAL TURBOPROP AIRCRAFT

It was a clear mid-morning over Nebraska, and I was just beginning my aerial application run on a field I knew well - one I'd sprayed many times before, even earlier that day. Focused on the GPS light bar mounted to the nose of my experimental turboprop, I was locked in on aligning my heading for the first pass.

As I leveled out, finally satisfied with the heading, my eyes caught something just beyond the light bar - several construction workers standing on a bridge at the far end of the field. I hadn't yet released any product, but instinctively hauled back on the stick to avoid startling them. From their perspective, I'm sure it looked like a hotdog buzz job. It wasn't.

Now behind schedule, I weighed my options. Skipping this field would throw off my whole day. So I decided to continue operations from the far side of the field, working back and forth on each swath, inching my way closer in hopes the workers would eventually clear out.

They didn't.

When I reached within 1,000 feet of their position, I called it.

I aborted the operation and flew back to base, still carrying a partial load. Minutes later, dispatch got a call from the bridge crew. Apparently, they "had to evacuate the area" and wanted to know what I was spraying - even though I hadn't applied anything near them.

Once I got confirmation that the crew had left, I returned and completed the job safely and without further incident.

Lessons Learned:

Familiarity is no substitute for situational awareness.

In agricultural aviation, it's easy to assume that a well-known field will be just as you left it - empty, quiet, safe. But people, vehicles, and unexpected hazards can appear without notice.

That morning, I let routine override procedure. Skipping a high recon cost me time and nearly caused a misunderstanding with a ground crew who had every reason to feel alarmed. They didn't see the GPS light bar I was watching, or know I wasn't spraying. All they saw was an aircraft bearing down.

In the future, no matter how well I know a location, I'll fly that recon pass. It's a simple step that ensures safety, builds trust, and keeps us all on the same page - above and below.

NOTES:

THE HIDDEN LIMB
ROBINSON R44 II

The sun was low, casting long shadows over the narrow fields outside New Florence. It was still hot, the kind of heat that lingers in the air long after the day has peaked. I'd been spraying all day and was setting up for another pass - routine work in a Robinson R44 II. Light, responsive, and familiar.

I was flying under Part 137. This was a job I knew well. With nearly 1,200 hours logged - and over 300 in this aircraft - I felt comfortable down low, navigating the tight corridors that come with this line of work. The R44 had been recently checked and felt sharp on the controls. I made one clean pass, finished the dispersal, and started climbing to reposition.

Everything was smooth - until it wasn't.

As I climbed out, threading between tree lines, I felt a sudden jolt. No warning. Just a hit. The tail had caught something. I didn't see it, but I knew right away - I'd clipped something with the tail rotor.

Almost instantly, the aircraft began yawing hard to the right. The pedal inputs didn't respond. The spin came on fast. I'd lost tail rotor effectiveness. The machine was no longer under control.

I didn't have altitude to work with. This wasn't the kind of emergency you recover from - it was already too low, and the ground was rushing up.

The helicopter hit hard. We came down in a spin, the impact collapsing the skids, buckling the tail boom, and crumpling parts of the fuselage. But somehow, it didn't catch fire. No fuel ignited. I stayed strapped in - shaken but intact.

I took a breath. The cabin was tilted and still. I unlatched and climbed out.

The helicopter was a mess. Bent, broken, grounded. But I was standing. My harness had done its job. My training had helped me stay focused. But none of that could change what had already happened.

The investigation didn't take long. No mechanical failures. No system issues. The controls had worked right up until the moment they couldn't. The conclusion was clear: I'd struck a dead tree limb. I never saw it. Maybe it blended with the shadows. Maybe I just missed it. But either way, it was there and I hit it.

The cause was listed as failure to maintain clearance during low-level operations. And they were right. That's what happened. It wasn't a failure of skill. It was a failure of awareness in that moment - one second where I didn't see what I needed to.

The R44 was severely damaged. It wouldn't fly again soon. But I walked away. That was something.

The follow-up was important. They reminded all of us in the industry about best practices - ground surveys, field mapping, updated charts.

And they reinforced the "Ferry Above Five" guideline - fly above 500 feet when transitioning. But in the middle of a spray run, 500 feet isn't an option. You fly low. That's the job.

The only buffer you have is awareness.

Lessons Learned:

Flying low leaves no margin for error. There's no radar for dead limbs. No terrain alert for a brittle branch. You're flying close to the edge, relying on sight, memory, and habit. And all it takes is one oversight - one obstacle you didn't spot.

That day, the adversary was a dead limb. Not moving. Not marked. Not flagged on any chart. Just a forgotten branch extending into a flight path I'd flown all day.

And that's the point. We don't crash because of the things we expect. We crash because of the ones we don't.

So build habits that help you survive the invisible:

- Fly the field before the job, even if you've flown it before.
- Update your own mental map with each pass.
- Watch for irregularities - broken lines, odd shadows, dead growth.
- And always assume something is waiting just outside your line of sight.

That day, I did a lot of things right. But the difference between a close call and a crash was one thing: a branch I didn't see.

It only takes one.

NOTES:

THE TURN TOO SOON
CESSNA A188

The heat shimmered off the pavement at the municipal strip that afternoon. It was just past 3 p.m., 34°C, and the wind was steady from the north-northeast at about 13 knots. I climbed into the Cessna A188, an older but sturdy tailwheel ag plane. I'd flown it a few times before - 22 hours in this make and model so far - but it still felt a bit new in my hands. My total time was 415 hours. I was building experience fast, and this was another routine Part 137 spray run.

The aircraft was built for this kind of flying. A Continental IO-520 pushing out 300 horsepower sat under the cowling. The mission was straightforward. The runway was dry. The weather was fine. I expected a clean departure.

I taxied into position on Runway 14, pushed the throttle forward, and started the roll.

Everything seemed normal - until it didn't.

As I reached rotation speed and eased the nose up, the aircraft veered hard to the right. It felt twitchy - "squirrelly," is how I'd later describe it. I added left rudder to correct, but the Cessna jerked the other way - hard left now. I tried to stabilize it, but the swing was fast, uncontrolled, and accelerating.

We were near the edge of the runway. There was no time to reset. I pulled the airplane into the air, just enough to get into ground effect, hoping I could ride the lift out and over the trees.

But I could see right away - I wasn't going to clear them.

I reached for the dump handle, trying to jettison the hopper load to buy a few more feet. I don't even remember if I got it all the way open. There simply wasn't enough time.

The trees came fast. We hit just off the runway.

The aircraft crashed hard. The left wing tore away. The right wing and lower fuselage absorbed the brunt of the force. We flipped, and the aircraft came to rest upside down, the mission over before it had even started.

I was still strapped in - shaken, but unhurt. Somehow, I had walked away. There was no fire, no explosion. Just quiet. Crumpled aluminum. Torn fabric. Flattened grass. I unbuckled, crawled out, and stood beside a wrecked airplane that I'd been flying just seconds earlier.

In the days after, the investigation began. They went through everything - controls, engine, linkages, structure. No malfunctions. No mechanical failure. Everything checked out.

What didn't check out was me.

The conditions that day created a crosswind on Runway 14 - 13 knots at 010°. For a pilot with limited tailwheel time, those numbers mattered more than I realized. Add in the effects of torque and P-factor during takeoff, and the loss of directional control made sense.

The aircraft didn't fail. I did.

The official finding was straightforward: loss of directional control during takeoff. The contributing factor? Crosswind. The cause? My inability to manage it in the moment it mattered most.

Lessons Learned:

Takeoff isn't just about horsepower. It's about control.

In taildraggers, control means directional control from the moment the wheels start turning. Rudder input has to be precise. Responsive. Balanced. And if the wind's off-center, even slightly, the margin disappears quickly.

On that takeoff, I didn't respect that margin.

Yes, I dumped the load. Yes, I reacted. But the error had already been made. I hadn't recognized the threat early enough. I didn't abort when I should have. I tried to fly out of a mistake that started on the ground - and I paid for it in bent metal.

So here's the takeaway: never underestimate the takeoff roll.

Know your crosswind components. Anticipate the aircraft's handling. If it feels off, stop early. Don't wait to see if it'll fix itself in the air.

You can't fix a blown takeoff with altitude you don't have.

Ag flying is about precision. And precision doesn't start in the air - it starts on the ground, with your feet on the rudder, ready for whatever the aircraft throws at you.

And when it throws more than expected, you'd better be ready to stop - before it's too late.

NOTES:

THE TURN THAT COST TOO MUCH
AYRES S2R-600

The sun had just crept above the horizon as I powered up for the first run of the day. By 7:06 a.m., I was airborne over the potato fields near Stanwood. Flying low and fast in an Ayres S2R-600, I leveled out at about 10 feet above the crops, holding steady at 130 mph. It was a familiar morning, and a familiar machine. The aircraft, built in the late '70s, had nearly 21,000 hours on the frame, but it was still doing the job.

I had just under 1,000 hours of flight time and had done plenty of these early spray runs for the company. The first pass was smooth - chemicals flowing, wind calm, everything steady. But something was nagging at me: the spray pressure.

To check it, I had to look left - way left. The analog gauge was mounted on the spray boom, outside the cockpit, about eight feet away. There was no quick glance. It required a full body turn, eyes away from the flight path, just to catch a clear reading.

And so I looked.

I turned my head to the 8 o'clock position, eyes focused on that small dial. Maybe it took a second. Maybe two. Long enough. Because in that moment, I wasn't flying the aircraft.

Then I felt an impact.

The main landing gear had struck the terrain. It slammed into the ground with enough force to send a shock through the entire airframe. I reacted immediately, pulling back hard, trying to recover. The aircraft surged upward, but I already knew the damage was done.

The left main gear had punched through the wing. I could see fuel starting to leak from the ruptured tank. There was no way I could keep flying.

I scanned for a place to put it down. Found a flat grass field, lined up, and set in for a forced landing. The aircraft touched down on its nose and forward fuselage. The propeller stopped. We slid to a halt.

I shut everything down, popped the harness, and walked away.

The investigation didn't take long. I told them exactly what happened. There were no mechanical issues. No systems failure. The aircraft had done everything right - until I stopped watching where I was going.

They confirmed it. The impact happened because I lost terrain awareness while checking a gauge. That analog pressure gauge on the boom had been there forever. You had to twist your body and turn your head to read it. At 10 feet off the ground and 130 mph, that's more than enough time to lose your margin.

The irony? The aircraft had a digital MVP-50T engine monitor installed - one that could've displayed spray pressure in the cockpit. But we'd reused an older boom system during the upgrade, and the digital display was never configured for pressure monitoring. The old analog gauge stayed. It was a compromise made in the hangar. One that caught up with me in the air.

I wasn't careless. I was doing my job.

But the job forced me to look where I couldn't afford to.

Lessons Learned:

Distraction doesn't always come from a phone or a call - it can come from a dial. And in ag flying, even a two-second glance is a lifetime. At low altitude, you don't get that time.

This wasn't about poor flying. It was about poor design. The system that required me to twist away from the horizon put me in a position to fail.

So here's what I take from it:

- Critical instruments need to be in the line of sight.
- If you have a digital system capable of showing key data - configure it.
- Don't settle for legacy setups if they compromise safety.
- And if you can't watch the field, don't fly the field.

I was lucky. I reacted quickly and walked away. But it never should've come to that.

When you're flying low, your eyes need to stay forward. Your hands on the controls. And your trust in systems that help - not hurt - you.

You can't fly what you can't see. So fix what distracts you - before it costs you more than a wing.

NOTES:

THE PUSH PAST THE POINT
RUMMAN/SCHWEIZER G-164D

The sun was still high when I taxied out at Richvale Airport that afternoon. It was June 7, 2024 - just after 5:00 p.m. - and the heat was oppressive. The ramp shimmered, 39 degrees Celsius and climbing. The scent of rice dust and jet fuel hung thick in the still air. It was the kind of day where every surface radiated warmth, and the horizon danced in the heat haze.

I've been flying for over four decades, with more than 14,000 hours logged - most of them in agricultural aircraft. That day, I was flying a Grumman/Schweizer G-164D, tailwheel configuration, built tough for fieldwork. It was powered by a PT6A-34AG - strong, dependable, and familiar.

The job: drop 2,400 to 2,500 pounds of rice seed over nearby paddies. I'd done this hundreds of times before.

I lined up on Runway 34. It's not a long runway - 2,200 feet, 50 feet wide. A bit tight, but usually enough. The wind, light and steady, was coming from 170 degrees. That gave me a tailwind component. Not ideal. But manageable. Or so I thought.

Throttle forward. The aircraft started its roll. The tail came up just past midfield. All my gauges looked good.

There were no warning lights, the engine temp was steady, torque holding where it should. But I could feel something wasn't quite right.

Acceleration was... sluggish.

The plane wasn't leaping off the line like I expected, especially with that load. I told myself it was just the heat. The weight. Maybe even the surface drag. I pushed the throttle all the way forward - "to the stops," as we say - and held it there. Full power. Give it everything.

But I was running out of runway. Fast.

By the time I hit the far end, I knew I wasn't going to make it. The gear struck the embankment beyond the asphalt with a thud that jolted the entire airframe. The left wing slammed into the ground. The load shifted forward with brutal force.

We skidded to a stop in a crumpled heap, nose down, wing twisted, seed spilled.

I just sat there for a second, listening to the silence.

Miraculously, I wasn't hurt. Not a scratch.

I climbed out and assessed the damage. The wing was torn open, the fuselage bent. The aircraft was done for the day - maybe for good. But I was alive, standing in the heat and dust, looking at a wreck that shouldn't have happened.

Later, I admitted what I already knew in my gut: I pushed it too far. That tailwind - only five knots - was enough to put me past the edge of performance.

The aircraft was fine. The engine ran flawlessly. There was no mechanical failure. The problem wasn't the airplane. It was the decision.

I chose to go, knowing the runway was short. Knowing the load was heavy. And knowing there was a tailwind. I relied on instinct and experience. Maybe a little pride.

But pride doesn't change physics.

Lessons Learned:

This wasn't a complex failure. There was no fire. No malfunction. Just an experienced pilot who made a bad call.

In ag flying, you're under constant pressure. Schedules. Weather. Fields waiting. But the truth is simple:

Heavy load + short runway + tailwind = trouble.

It doesn't matter how many hours you have. It doesn't matter how many times you've "gotten away with it." Margins are margins - and tailwinds eat them fast.

Five knots may not sound like much, but on takeoff, it can add 10% or more to your ground roll. Combine that with a high density altitude and a full hopper, and you're already behind the curve.

The lesson I took with me that day?

- Trust your numbers more than your feelings.
- Plan for performance, not for hope.
- Abort if acceleration feels off. Don't wait to see how it ends.

There's no glory in pressing on past the point of safety. And there's no shame in turning back. That day, I walked away. The aircraft didn't. But next time? There might not be a second chance.

So if you're ever tempted to push through, ask yourself one thing:

Would I make the same call with a student in the seat beside me?

If the answer's no, then don't.

NOTES:

ONE LEVER TOO FAR
AIR TRACTOR AT-502B

The sun had just climbed high enough to burn away the last of the morning haze when I took off from the dirt strip near Corcoran, California. It was early October, and the cotton fields stretched in neat rows, waiting for their final treatment. The skies were clear, the breeze light out of the south, and the temperature hovered at 27°C - perfect for flying.

I've flown more than 8,000 hours, with nearly 2,000 in the Air Tractor AT-502B. N243LA was as familiar to me as my own hands - tough, responsive, dependable. This was routine work under Part 137: quick application runs over fields I knew well, flying from a 5,000-foot private strip, 80 feet wide and more than enough room to operate safely.

At 10:25 a.m., I launched into the air. The aircraft responded beautifully. It was a short run - fly the load, circle the farmland, and return. Everything went according to plan. The hopper emptied clean, the wind behaved, and I set up for final with the strip directly ahead.

Then, in the final turn from base to final, I glanced down for a last instrument check - and my heart dropped.

The fuel condition lever was in the cutoff position.

It hit me like a slap. That lever regulates the flow of fuel to the turbine. In cutoff, there's no fuel - and no engine. Sure enough, I felt the engine spool down beneath me. I was losing power. Fast.

There wasn't time to think, only act. I slammed all three levers - power, prop, and condition - full forward. I hoped that somehow the fuel would catch, the turbine would relight, and the engine would roar back to life.

But nothing happened.

There was no surge. No recovery. Just silence, and a rapidly approaching field.

I aimed the nose at the only option in front of me: a cotton field. There were no obstacles, just soft earth and long rows.

The landing was hard. The gear struck first, bouncing once. Then the nose dug in. The left wing took the brunt of it, folding slightly on impact. But the aircraft stopped upright. And I was still alive.

No fire. No fuel leak. No explosion. Just silence and the realization that I had put the lever in the wrong place. I unstrapped, stepped out, and stood there beside the wreck, heart pounding, thinking how close I'd just come.

When I spoke to investigators later, I didn't sugarcoat it. The aircraft had no mechanical issues. The engine was fine. Fuel was clean. The system had done exactly what it was supposed to do.

The failure was mine.

At some point - maybe during a quick flow check, maybe in turbulence or distraction - I'd moved the condition lever to cutoff. It didn't take much. A few centimeters. One lever. That's all it took to turn a routine return into a forced off-field landing.

Lessons Learned:

This wasn't about maintenance, weather, or hidden hazards. It was about the brutal simplicity of human error.

When you're alone in the cockpit, especially in fast-paced agricultural ops, every motion matters. Every lever and switch must be where it should be, every time.

The condition lever is one of the most critical controls on a turbine. It's the gatekeeper between flight and flameout. And it's not enough to check it once. You have to build habits that check it again - and again.

What nearly brought me down wasn't a mechanical failure or poor judgment. It was a simple oversight in a familiar environment. The kind of mistake that can happen to anyone, no matter how many hours are in your logbook.

So if there's one takeaway from my story, it's this:

- Respect the routine. Don't rush through it.
- Make cockpit checks sacred - especially on approach.
- And always, always confirm that the condition lever is where it belongs.

That day, I was lucky. I got to walk away. The airplane didn't. But I got something far more valuable than the aircraft: the chance to fly again - and to remind others just how fast one small error can undo a thousand safe flights.

Check the lever. Then check it again.

NOTES:

ANGLE OF NO RETURN
AYRES S2R-T34

The evening heat clung to everything as I prepared for what should've been another routine run. I was flying a heavy Ayres S2R-T34 ag plane - rugged and built for the work. I had over 7,000 hours logged and held multiple ratings. That day, I had 300 gallons of water on board, fuel in the tanks, and a short strip ahead of me with a 15-knot breeze down the nose.

Everything checked out. I lined up for takeoff and pushed the throttle forward.

Immediately, something felt off. The acceleration just wasn't there. I was rolling, but the end of the runway was coming fast. I hit the emergency dump and shed the full load of water. That helped a little, but it still wasn't enough. I was at the end of the strip when the wheels finally lifted.

I was flying - but only just.

Trees loomed ahead. I didn't have altitude, so I pulled back hard, trying to muscle the plane up. The nose rose steeply. I could feel it clawing upward - and then it happened. The airspeed bled away, the lift disappeared, and I knew right then I was out of options.

The aircraft stalled.

There was no time to recover. We pitched forward and hit the field just past the runway. The impact was violent. Metal crumpled, fiberglass cracked, and control surfaces took the worst of it. But I climbed out without a scratch.

The airplane wasn't so lucky.

I sat beside the wreck for a while, thinking through every moment. There hadn't been any warning lights. The engine had run fine. The dump system did its job. Everything that was supposed to work had worked.

Later, I figured the problem was density altitude. The temperature was in the mid-30s Celsius, and although the elevation wasn't extreme, the performance definitely was. It had fooled me. And when it came time to climb, I asked the wing to do more than it could give.

I didn't make that mistake out of panic - I made it trying to do the job. But that doesn't change the outcome.

Investigators later confirmed what I already knew: I had exceeded the critical angle of attack. The stall was aerodynamic - pure and simple. There were no mechanical failures. No gusts. Just a steep pitch, low airspeed, and no altitude to fix it.

Lessons Learned:

This one taught me what books and briefings sometimes don't drive home: in ag flying, energy management isn't optional - it's survival.

Here's what I learned:

- A high pitch doesn't mean a high climb. You can't just point it up and expect it to go.
- Dumping the load helps, but it doesn't solve poor energy management.
- Density altitude can eat into performance faster than you expect - even when you're close to sea level.

- If the climb feels off, trust your gut. Abort and reset if you have any doubt.

I reacted quickly to the underperformance, but I made the mistake of trying to climb over terrain by pulling harder. The wing said no. And when that happened, gravity answered.

I survived that day. The aircraft didn't. But I walked away with the clearest understanding yet of what it means to earn your altitude.

Airspeed first. Every time. Don't ask your wing to do more than physics will allow. Because when it stops flying, all you've got left is whatever ground lies ahead.

NOTES:

DRAG AT THE WRONG TIME
AIR TRACTOR AT-502B

The heat in Bakersfield was rising fast as I prepared for my fourth spray run of the day. I'd flown plenty of missions like this one. The aircraft, an Air Tractor AT-502B was fueled with 50 gallons and loaded with 400 pounds of fertilizer. Nothing unusual.

The strip was tight - 2,500 feet long, just 50 feet wide. Not generous, but I'd operated off it more times than I could count. With nearly 6,000 hours in my logbook, over 700 in this model, I felt confident.

I set the flaps to 20°, ran through my checks, and lined up. The sky was clear, the wind mild, and I was focused on getting this last job done before the heat peaked.

With brakes held, I pushed the throttle to the firewall. The turbine responded with a roar, and I released the brakes. The aircraft surged forward. It felt fine - until I hit about 1,300 feet down the runway. I was still stuck to the ground.

I eased back on the stick and managed to get airborne, but just barely. We floated into ground effect - maybe 10 feet up - but I couldn't get any higher. The aircraft simply refused to climb.

Then instinct took over.

I dropped the flaps to full, thinking the extra surface might give me lift. I knew it was a risk, but I didn't have time to reason it out. Trees loomed ahead. I needed height.

But the full flaps didn't lift me - they dragged me.

The aircraft started to sag. The extra drag stole what little momentum I had. I touched back down near the far end of the strip, too fast and too far down to stop.

We rolled off the end and into the trees.

Branches shredded the wings. The aircraft jolted to a violent stop. Everything went still. I was bruised, shaken - but alive. The airplane wasn't so lucky.

After climbing out and walking the scene, I replayed every second. No alarms. No system faults. The engine had pulled. The weather hadn't shifted. Nothing had failed - except my decision-making.

Later, I reported the accident. I mentioned a possible mechanical fault, but I didn't include everything in my written narrative. Still, the truth came out in the investigation.

There had been no malfunction. The aircraft performed as it was built to. The mistake was mine.

The investigators pointed to one key factor: full flaps in a short-field takeoff. It's right there in the manual - don't do it. The drag from full flaps can destroy takeoff performance, and that's exactly what happened here.

In that moment of desperation, I went against the book. I thought I was helping the aircraft climb - but I took away its ability to accelerate and lift at the worst possible time.

Lessons Learned:

Procedural discipline matters - especially when you're operating near the margins.

Here's what I took away:

- Full flaps and short fields don't mix. They generate drag, not climb.
- Stick to the aircraft manual. It's written from lessons already learned.
- If climb performance is weak, don't guess. Evaluate based on what you know, not what you feel.
- Resist the urge to improvise unless it's truly necessary - and backed by solid understanding.

I walked away with cuts and a bruised ego. But the lesson is permanent.

Don't fight physics with hope. If the plane's not climbing, adding drag isn't the answer. Know your aircraft, fly by the book, and when the margins are tight - be conservative.

That mistake cost an aircraft. It could've cost more.

I'll never forget the sound of those trees - or the silence that followed.

NOTES:

ONE DRIFT TOO FAR
ROCKWELL S-2R

It was a calm morning in June when I climbed into the cockpit of my ag plane for what should've been another routine run. I'd flown this aircraft hundreds of times - a Rockwell S-2R. With over 2,000 hours total time, nearly 900 in this type, I knew its quirks and temperament well.

Conditions were ideal. Clear skies, a light breeze from the southwest, and no obstructions in the area. The aircraft had just come through a fresh inspection. Its turboprop engine was delivering full power, 1,000 horses ready to work. The strip I was using was tight - about 3,250 feet long and just 35 feet wide. Mostly concrete, part dirt. Narrow, sure, but familiar.

The job was local - just a quick departure, apply product, and return. I'd done this kind of mission many times. Nothing about it raised a red flag.

I taxied into position, ran the checks, and began the takeoff roll. That's when I felt it - the subtle, almost imperceptible drift to the right. Not unusual in a tailwheel aircraft, especially with a light crosswind. I tapped left rudder. No change. I pressed harder. Still, the aircraft kept pulling right.

I didn't panic. I figured I could correct it, ride it out. But the drift got worse. The airplane rolled off the strip and over a small rise. I wasn't aligned with the takeoff path anymore. But instead of aborting, I made a snap decision: I shoved the power to full and tried to fly out of it.

The aircraft got airborne - just barely. But I could feel it wasn't right. The nose pitched up into an odd attitude. Unstable. Nose-high. The climb didn't feel safe. I pushed forward on the yoke to level out and get the nose down.

Too late.

We came down hard. The impact tore off the left main gear and spreader bar. We skidded and rotated, eventually coming to rest across the runway, battered but upright.

The left wing and aileron were wrecked. Debris scattered everywhere. But I was okay. Shaken, but unhurt.

After everything stopped, I shut down the engine and climbed out. I looked over the wreckage and started piecing it all together. There'd been no mechanical issues. No surprises from the aircraft. It was all me - directional control lost on takeoff.

I later reported what happened and confirmed everything on the aircraft worked as designed. It was pilot error. I tried to salvage a takeoff after we were already in a bad position. And it didn't work.

Investigators didn't come to the scene. They used my account, documentation, and photos to figure out what happened. It wasn't complicated. I had the experience. The aircraft had the power. The weather wasn't a factor. But the strip was narrow, and I made a poor call when the drift started.

I didn't stop it early. I didn't abort. I tried to force a solution instead of making the safer call.

And that turned a manageable drift into a full-blown accident.

Lessons Learned:

Ground handling in tailwheel aircraft isn't forgiving. Especially on tight strips with heavy loads. A little drift can become a big problem fast.

Here's what I learned:

- Correct drift immediately. Don't wait to see if it sorts itself out.
- If you lose directional control on the roll, abort. Don't try to save it in the air.
- Tailwheel aircraft magnify small mistakes during ground roll. The window for error is tiny.
- Trying to fly out of a botched takeoff might feel right - but unless it's backed by airspeed and alignment, it's a dangerous gamble.

My instinct was to keep going. I figured I could power through it. But I was wrong. And that decision cost me the airplane.

I walked away. Others in similar situations haven't been so lucky.

In ag flying, everything happens fast and close to the edge. If the airplane starts to get away from you, stop. Don't hope. Don't push. Make the call early - and live to fly again.

NOTES:

THE TANK RAN DRY
BELL UH-1B

It was a clear March day, perfect weather for ag flying. I was at the controls of a Bell UH-1B helicopter, an old Vietnam-era machine now repurposed for agricultural work. I've been flying for decades, and by that point, I had more than 21,000 hours in the logbook. It was another day of application runs in northern California, and everything had gone smoothly since I'd lifted off that morning.

The helicopter had been performing flawlessly all day. The air was calm. Visibility stretched for miles. Run after run, I worked the fields, moving low and fast, covering ground with the kind of rhythm that only comes from thousands of hours in the seat.

By midafternoon, I was on what I thought would be the last run of the day. That's when the engine started to spool down. It wasn't sudden or dramatic - just a noticeable drop in turbine whine. I knew immediately that something was wrong.

There wasn't time to troubleshoot.

No time to scan every gauge or pull out a checklist. At low altitude and low speed, you don't get to ask why - you just react. I scanned for a place to set down and spotted a narrow road nearby. It was my best shot.

I turned toward it, trying to hold a stable glide path. But without power, I was dropping too fast. The road wasn't going to happen. I aimed for the flat terrain just short of it and tried to cushion the landing as best I could.

The skids hit first. The helicopter bounced hard. The airframe flexed, the blades bent dangerously, and I braced for what I thought might be the end.

But somehow, it held together. I was shaken but uninjured.

Once everything stopped moving, I shut down what was left and climbed out. No fire. No secondary damage. Just the deep quiet of a field and a broken helicopter.

Later, when I was asked what happened, the answer was simple: the aircraft had run out of fuel.

There had been no leaks. No mechanical faults. The systems had all worked as they should. The helicopter had passed its annual inspection just weeks earlier. The engine was strong and capable. But I hadn't planned the fuel load properly, and I'd run it too close to empty.

It was on me.

I admitted it during the interview. There was no getting around it. I'd pushed the last leg too far without topping off. The aircraft had done its job. I hadn't done mine.

Looking back, it's still hard to believe. With all those hours, all that experience, I still made the most basic mistake a pilot can make.

Lessons Learned:

Fuel exhaustion remains one of aviation's most avoidable failures - and yet it continues to happen. It happened to me.

Experience doesn't override the laws of fuel consumption. Gauges don't care how many flight hours you've logged. If you don't have the fuel, you're not going to make it.

Here's what I learned - what every pilot should remember:

- Fuel checks must be routine, every time, no matter how well the day is going.
- Don't stretch your reserves to squeeze in one more job. One more run can be the one that ends everything.
- Plan with a buffer and stick to it. Fatigue, pressure, and optimism are not a fuel management strategy.
- It doesn't matter how flawless the machine is if you don't feed it.

That helicopter had done everything I asked of it. The only failure was mine - failing to check, failing to plan, and failing to respect the limits.

I walked away that day, but I left behind a wrecked aircraft and a scar on my own judgment. It was avoidable. Completely avoidable.

Next time, for me - or for you - I hope the lesson is remembered before the tank runs dry.

<u>NOTES:</u>

TOO CLOSE FOR COMFORT
AIR TRACTOR AT-402A

That morning, I was flying my Air Tractor AT-402A over Alabama, finishing up a short aerial application run. I was operating off a narrow 2,500-foot strip - just 24 feet wide - with tall wheat flanking both sides. The field was familiar. I'd flown out of it many times before.

The skies were clear, the winds light, and the aircraft - tailwheel, 550 horsepower - was performing perfectly. I had over 2,000 hours logged, 800 in this model. I knew how to handle tight strips. But something about that runway looked different on return. The wheat seemed taller. The margins tighter.

Still, I felt confident.

Lining up for landing, I reminded myself to stay centered. With crops that close, there wasn't room to drift. I touched down smoothly. Everything seemed under control - until the rollout.

I began to drift slightly right. Just a foot or two. That's all it took.

Suddenly, the right spray boom caught the tops of the wheat. It was a light snag - but enough to pull the aircraft further off course.

I felt the plane tug sideways. The right wing dipped. Before I could react, I was off the strip and into the field.

The airplane tore through the wheat. The left wing and tail structure crumpled as the plane skidded through the dense crop. Dust and broken stalks flew around me. Then everything stopped.

I sat still. The aircraft was upright. I was unharmed.

Climbing out, I did a quick assessment. The aircraft was badly damaged. The field looked like a harvester had passed through, but I was okay.

Later, I reviewed everything I'd done. There had been no mechanical issues. No gusts. No problems with the aircraft at all.

I'd simply underestimated the clearance. I hadn't fully factored in how much that wheat had grown since the last time I landed there. I'd relied on habit, not on fresh judgment.

Investigators came to the same conclusion. Everything on the airplane functioned properly. There were no signs of malfunction. The problem had been with my decision to land on a strip where the surrounding crop left almost no margin for error.

It was a reminder that fields change. Familiar places evolve.

Lessons Learned:

Agricultural aviation demands more than just skill - it requires attention to detail, even when things seem routine.

Here's what I learned:

- Nature doesn't wait. Crops grow fast. What looked clear yesterday may not be today.
- A 24-foot-wide strip offers no forgiveness. If your wheels are centered but your boom isn't, you're vulnerable.
- Verify lateral clearance every time. Especially in late-season growth, clearance assumptions can turn dangerous.

- Just because the field is familiar doesn't mean it's safe. Familiarity breeds comfort - but also risk.
- Ag pilots operate with razor-thin margins. A few inches of crop can take out a boom, a wing, or worse.

I was fortunate. I walked away. The airplane didn't flip. It didn't burn. But I knew I'd been on the edge.

This flight reminded me that our environment isn't static. It shifts with every season, every storm, every crop cycle.

The field didn't fail me. The airplane didn't fail me. What failed was my assumption that everything would be just like it had been before.

In ag flying, comfort is the enemy of precision. Even a slight drift - on a narrow strip flanked by tall crops - can turn a normal landing into a wreck.

Measure your margins. Respect the crop. Reassess every time.

NOTES:

THE POWER THAT WASN'T
AIR TRACTOR AT-502

The morning was perfect - calm, clear, and ideal for flying over North Dakota farmland. I'd been flying my Air Tractor AT-502 for a local spray job, and everything was going smoothly. The aircraft had been inspected just days before, and it was flying beautifully.

After finishing the application run, I began the turn back toward base. That's when I felt it - something wasn't right.

The engine tone shifted, and the torque gauge dropped suddenly. It caught my attention immediately. I pushed the power lever forward, expecting the usual surge. But nothing changed. No power increase. No response.

Fields and obstacles surrounded me, but there was a straight stretch of county road ahead. I didn't think - I just committed. I was going in for a forced landing.

I touched down hard. The tires hit the road, bounced once, and then I floated briefly before slamming down again. That second landing was worse. The aircraft veered, left the road, and tore through a ditch before coming to a stop in the field beyond.

I was unhurt. The plane wasn't so lucky. The gear was wrecked, and the prop and engine had taken damage.

Over the next few days, something started gnawing at me.

I pulled the GPS data - speed, altitude, track - and started reviewing it frame by frame. The numbers didn't support what I remembered. There was no drop in airspeed. No real altitude loss before touchdown. The engine hadn't failed the way I'd thought.

Then came the mechanical checks. The fuel control unit and fuel pump tested fine. There was some minor fluorocarbon debris near the Py orifice, probably from an O-ring, but nothing that would've blocked flow or caused failure. Everything worked.

That's when it hit me.

I had misread the situation.

I updated my report, owning the truth: "My memory of the event in that moment is patchy, scrambled, and likely inaccurate." The engine hadn't quit. But I'd believed it had - and that belief shaped everything that followed.

I'd reacted to a perception, not a fact.

The final assessment was tough to read but fair. The cause: a hard landing followed by a loss of control. Supporting factors: pilot error, misjudgment, and directional control failure. No system malfunction. Just a string of actions built on a faulty assumption.

Lessons Learned:

This experience taught me something I'll never forget: perception under pressure can betray you. Your senses might scream that something's wrong, but unless it's backed by the instruments, it could be a phantom.

- Trust the data. One gauge may twitch - but cross-check the rest.
- Stay calm when things feel off. Reacting too fast to one input can compound the problem.

- Debrief with honesty. I had to confront the fact that I got it wrong. And that was humbling.
- Experience helps - but it doesn't immunize you from error.

The plane was damaged, yes. But the real damage could've come from doubling down on the wrong narrative. Fortunately, I had the tools - and the team - to work through the facts.

Sometimes, the lesson is this: what you think is happening isn't what's happening.

In aviation, that difference can cost you everything.

NOTES:

THE RIDGE I DIDN'T SEE
PIPER PA-25

It was early afternoon in August, and the sky over northeast Nebraska was pristine - deep blue, not a single cloud in sight. The kind of weather that makes you believe nothing can go wrong. I was flying a Piper PA-25, working a field I'd flown over dozens of times before. Another agricultural application job, another swath of land to treat, another day in the cockpit.

I'd already made several passes, hugging the terrain just like we're trained to. Low-level flying in a taildragger - it's an art form. Each field is a puzzle, every approach slightly different. But something about this pass would prove different, something I didn't catch until it was too late.

I lined up again, dropped down into the familiar dance of application height, and adjusted for the next run. Everything seemed routine. The crops below looked uniform. But they weren't.

What I hadn't noticed was a subtle elevation change - a slight rise in the slope of the crops on the far side of the field. It wasn't dramatic. Just a gradual increase. Enough to change everything at the speed and altitude I was flying.

My gear caught first. A sudden jolt.

I instinctively added power, trying to claw the aircraft free, but it was no use. The drag of the entangled crops robbed me of energy, and in seconds, the Piper settled into the field and struck the ground. The impact was firm but not violent. The plane came to rest upright, but the left wing was torn and the empennage twisted. I sat there for a moment, catching my breath in the stillness.

There was no fire. No explosion. Just a damaged aircraft and the unmistakable thud of reality settling in.

Later, I walked the field, retracing the pass. That was when the rise became clear. A gentle slope, camouflaged by the uniform height of the crops until you were nearly on top of it. I'd misread the terrain. Or rather - I hadn't read it at all.

No mechanical issues. No system failures. Just a simple lapse in awareness.

Lessons Learned:

In agricultural flying, the greatest threats often aren't dramatic - they're subtle.

The field doesn't always shout. Sometimes it whispers. A gentle slope. A subtle ridge. A minor change in the shape of the land that, at high speed and low altitude, becomes the difference between a clean pass and an accident.

We pride ourselves on reading terrain, knowing our fields, anticipating the environment. But familiarity can lull you into missing the smallest shifts. That day, I let routine and rhythm override observation. I assumed the field was what it had always been, and in this job, assumption is as dangerous as inattention.

Flying low demands more than precision - it demands humility. A willingness to scan each field as if it's the first time. Because it only takes one pass to remind you: the land might be quiet, but it never stops speaking. And if you stop listening, it'll let you know.

NOTES:

THE TAKEOFF THAT WOULDN'T LET GO
AIR TRACTOR AT-502

It was just after sunrise in August. The air was still cool, the sky clear, and the wind calm - or so I thought. I was preparing for the first spray flight of the day from a narrow strip in Halstad, a 2,500-foot asphalt ribbon bordered by soybean fields and Minnesota silence.

The Air Tractor AT-502 was loaded heavy: around 100 gallons of jet fuel and more than 400 gallons of product. I'd flown this aircraft for years - nearly 1,700 hours in this make and model, and more than 9,000 total. This was supposed to be routine. But ag flying has a way of humbling even the most seasoned hands.

I lined up on the south end of the strip, pushed the throttle forward, and started the takeoff roll. I had one notch of flaps in and a full load behind me. I knew there was a tailwind, maybe six knots - not ideal, but manageable. Or so I told myself.

With about 300 feet of runway left, I rotated. In hindsight, I rotated too soon. She got airborne, but not really. We were stuck in ground effect, skimming just above the surface but not climbing. I had no room left. I reached for the dump lever and began jettisoning product. But we were already past the threshold, and the soybean field ahead was closing in.

For about 200 yards, we hung in the air - barely. And then the weight, the lack of lift, and the wind pushed back harder than I could fly through. We sank into the field, the aircraft settling with a jolt that echoed through the frame. The wings tore as they hit the beans, the empennage crumpled, and the fuselage buckled beneath me.

I sat there in the cockpit, unhurt, staring out at the early morning light and the wreckage around me. The silence was absolute.

Later, when I retraced what happened, it became painfully clear. The aircraft was mechanically fine - no failures, no surprises from the engine or airframe. The failure was mine. I tried to take off heavy, into a tailwind, on a short strip, and I rotated before the aircraft was ready to fly. I pushed the machine beyond its performance envelope - and it pushed back.

Lessons Learned:

Every airplane has its own limits - but it's the pilot's judgment that defines whether those limits are respected or crossed.

Tailwinds reduce climb performance. Heavy loads stretch margins. And ground effect can deceive you into thinking you're flying when really, you're just skimming along the edge of control.

I knew better. I knew the strip, the load, the wind. I made a choice anyway - one that cut too close.

In ag flying, we often operate on the edge - short fields, full hoppers, demanding schedules. But no job is worth gambling on performance. The laws of physics don't care about your confidence or your experience.

Takeoff is a commitment. Make sure the conditions are ready to commit with you. Because once you lift off - if you're not ready to climb, you won't stay airborne for long.

NOTES:

THE SPIN I COULDN'T STOP
ROBINSON R22 BETA

It was a hot, windy day in July - the kind of day where everything feels just a little more unpredictable. I was flying a Robinson R22 Beta, my first day out on aerial spray operations. I'd trained hard for this, logged hours in the right seat, and now here I was - on the job, rotor turning, confidence high, but nerves quietly present.

Hovering low, maybe ten feet above a rice field, I was transitioning from a west to east heading. That's when the wind hit - harder and gustier than I'd experienced in training. I tried to stay ahead of it, but almost instantly, I felt control start to slip away.

The tail started to swing. Clockwise. Fast.

I fought it with everything I had - pedals, cyclic, mental checklists flying through my head - but the wind was beating me up. My training hadn't fully prepared me for this kind of gusty assault, and I felt it. The spin accelerated. I couldn't stop it.

Seconds later, the helicopter slammed into the field.

The impact tore the tail boom, mangled the rotor blades, and crumpled the machine. Somehow, I walked away with only minor injuries. Bruised pride, sure. But physically intact.

The R22 had no mechanical failures.

The systems were fine. The maintenance was up to date. The machine had been ready for the job.

But I wasn't.

Lessons Learned:

First days can be the most dangerous - not because of what you don't know, but because of what you haven't experienced.

Wind is a living thing. It doesn't care about your logbook or your optimism. Gusts can overwhelm even seasoned pilots when they come from the wrong angle at the wrong moment.

This wasn't about reaction time - it was about preparation. Training in calm or moderate conditions didn't build the muscle memory I needed for sudden, aggressive gusts close to the ground.

Now I know: low altitude, tight maneuvers, and gusty wind is a combination that demands respect. And that confidence in the aircraft must always be matched by humility in yourself.

Because when the tail starts to swing, and the wind decides it's in charge - there's no time left to wish you'd trained for this.

Train harder than you think you need to. Because the real test doesn't come when you're ready - it comes when you're barely hanging on.

NOTES:

THE BETA THAT BIT
AIR TRACTOR AT-502

It was June. The heat was up, the altitude was high, and I had just taken off in the AT-502 with a heavy load: 660 pounds of fuel and over 3,000 pounds of chemical in the hopper.

The job was straightforward - a pasture not far from the strip. But shortly after arriving, I noticed something wrong. The spray system wasn't working right. No use pushing through. I decided to return to the airport and land to troubleshoot.

I came in high and fast - about 125 mph on final. Half flaps down, tailwheel locked. The plan was to touch down in the first quarter of the dry concrete runway and use beta to slow the aircraft.

As the mains settled and the tailwheel touched, I initiated beta mode with the propeller. Too hard and too fast.

The aircraft veered right immediately. It wasn't subtle. The reversal of thrust caught with too much force while the plane was still unstable. I tried to correct with rudder and brake, but I was already off balance. We were off the centerline in seconds.

The left gear caught the edge of a ditch beside the runway and sheared off.

The airplane lurched, then slid sideways into the grass field. When we stopped, we were upright, but the damage was done - the left wing crushed, the fuselage twisted.

I walked away without a scratch. But I walked away knowing I'd made a bad call.

Later, we inspected everything. The aircraft was sound. No mechanical failure. No engine issue. The malfunction in the spray system was traced to debris in the lines - simple blockage, easily fixed on the ground.

But none of that caused the accident. I did.

Lessons Learned:

A taildragger with a full load doesn't forgive rushed decisions - especially not at 6,100 feet density altitude.

Beta mode is a powerful tool. Used right, it shortens landing rolls and helps manage heavy landings. Used wrong - too early, too aggressive - and it becomes a hammer when you need a scalpel.

The real problem wasn't just the beta - it was rust. I hadn't done a full-load landing in over two years. The skills were there in theory, but not in muscle memory. In high-density altitude, small mistakes grow teeth. There's less room to work, less time to think, and every input counts.

I knew better. But I rushed the reversal, trying to make the aircraft behave on my terms instead of respecting its limits.

Now, I rehearse full-load landings even when I don't have to. I remind myself that when the aircraft is heavy, the runway is short, and the air is thin - there's no such thing as a routine landing.

Every landing starts with restraint. Because what you do in the first few seconds after touchdown decides whether you taxi in - or get towed out.

NOTES:

THE TURN THAT TOOK ME DOWN
GRUMMAN G-164B AG CAT

It was May, late morning, and already warm enough that the air had a little weight to it. I was flying out of a private strip near Light, Arkansas - a dirt and turf surface I knew well. That day I was hauling fertilizer, not quite at max gross, but still heavy. Around 320 pounds shy of the limit. Conditions were good, visibility was clear, and I'd done it all before.

I lined up, ran through my checks, and pushed the throttle forward. The engine roared to life, the tail lifted, and soon enough, I was airborne. Everything felt right. But right after takeoff, I made a choice I'll never forget.

I turned early.

Maybe it was habit, or comfort. I'd flown that field so many times before that it just felt natural to start turning back toward the application area, even though I hadn't yet built up solid airspeed or altitude. Just a little premature - just a little eager to get the job going.

The aircraft responded at first, but I could feel it straining. I asked for more lift, and it had nothing left to give. The wings weren't biting hard enough. The airplane sagged beneath me, the turn bleeding off what little margin I had.

Then we settled.

The Ag Cat dropped into the ground and struck a levee. There was no time to recover, no time to correct. One second I was flying, the next I was flipping through the air.

When it stopped, I was upside down in a field, staring at cracked glass and twisted metal. My harness had done its job. I was bruised, but I was alive.

The damage was massive. Wings - shredded. Tail - crushed. The fuselage twisted and scarred. The airplane had given everything it had, and it still wasn't enough.

When we went through the post-crash report, nothing stood out. No mechanical issues, no engine trouble. Everything on that airplane was working. The conditions - about 2,000 feet density altitude - weren't ideal, but not extreme.

I had power. I had runway. I had options.

And yet I chose to turn too soon.

Lessons Learned:

Takeoff isn't just about getting off the ground - it's about knowing when you've truly earned the right to climb and maneuver.

When the airplane is heavy and the air is thin, every pound and every foot counts. A turn that feels routine when you're empty becomes a gamble when you're loaded.

In my case, that gamble flipped a perfectly good airplane upside down in a field.

It's easy to get comfortable, especially when you've flown thousands of hours in the same type. But comfort doesn't equal immunity. The airplane still obeys the laws of physics - not your familiarity.

Now I make sure I've got the speed, the altitude, and the room to climb clean before I even think about turning, no matter how many times I've flown the field.

Because in ag flying, the margin for error is often just one premature decision away from disaster.

Wait the extra second.

Fly the extra hundred feet.

It might just be the difference between making a pass, and becoming a story.

NOTES:

BOUNCE, FLARE, AND THE DITCH
AIR TRACTOR AT-400A

The heat of June was already thick in the air as I lined up on final for a routine return to the private strip in Pemberton Township. It was familiar territory, just another landing after a long spray day. The runway was a rough turf surface, but I'd put it down there plenty of times before without issue. I came in clean, gear down, locked in on the numbers.

But I was too fast and too high.

I started the flare late, expecting I could smooth it out like always. But the airplane didn't settle - it bounced. Not just a hop, but a full slap back into the air. I tried to correct, tried to ride it out, hoping I could recover. I added a little throttle, trying to find the cushion and reestablish control.

It bounced again - harder this time.

Now I knew I had to go around. I pushed in power, but the angle of attack was too steep, and I was already behind the curve. Instead of climbing out clean, the aircraft veered left. I fought it, but the correction came too late. We were off the runway and into the ditch.

The landing gear collapsed instantly.

I felt the airframe crunch beneath me, the nose dipping hard as the ditch swallowed the front wheel. Dust and dirt filled the cockpit. When everything stopped moving, I sat there for a moment, stunned.

I was unhurt, but the aircraft wasn't.

The main gear was torn off, the fuselage twisted and scarred from the impact. The damage was serious. I shut everything down and climbed out, already replaying the seconds before the crash in my head.

There were no mechanical issues. Nothing failed on the airplane. This one was on me.

Lessons Learned:

A go-around isn't a backup plan - it's a primary decision, and it has to come early.

That day I underestimated the bounce and overestimated my ability to recover. I thought I could salvage it. I waited one second too long, and by the time I made the choice to go around, I didn't have the airspeed, the space, or the control I needed.

In taildraggers, especially on rough strips, timing the flare is everything. You can't afford to be behind the airplane. And when something feels off, hesitation is the real danger.

I learned that a late flare and a late go-around can stack against you quickly. Next time, I'll act on instinct - early and committed - because the window for recovery closes fast, and the ditch is always closer than you think.

NOTES:

SLIPPED AND FLIPPED
AIR TRACTOR AT-401B

It was early in the morning, and the dew was still clinging to the grass when I lined up for landing after another aerial application run. I'd flown into this field plenty of times, and on paper, everything looked good. Light winds. Visual conditions. Nothing tricky in the setup. But it's the small decisions - the ones that seem insignificant in the moment - that catch you out.

I was coming in on runway 31. The wind was light but from the southeast - about 4 knots on the tail. I figured it wouldn't make much difference. I was high and a bit hot on final but thought I could ride it out. The grass was wet, but I'd handled slick strips before.

I flared too fast, touched down long, and the wheels didn't bite.

Instead of slowing down, I was floating on the dew-slick turf, tires slipping just enough to keep me from getting full control. The end of the runway rushed up. I hit the brakes harder, trying to salvage it, but it was no use.

We overran the end, the main gear hit uneven ground, and the nose dropped. In a blink, we flipped, nose over tail, and came to a stop upside down.

I hung in the straps, heart pounding, before slowly unbuckling and crawling out. The rudder and horizontal stabilizer were wrecked. The airframe was bent and bruised. I was lucky to be unhurt.

There was nothing wrong with the airplane. No mechanical failure. No surprises. Just a pilot who made a poor decision under familiar conditions.

I'd underestimated the tailwind. I'd overestimated the braking ability of wet grass. And I'd let my approach speed get away from me. Each of those alone might have been recoverable. Together, they flipped the airplane.

Lessons Learned:

Tailwinds, wet grass, and optimism don't mix.

It's easy to dismiss a light tailwind or a slick surface as minor factors - especially when you've got a field you know and a routine you trust. But small risks stack fast. A couple of knots, a few extra feet per second, and a little less grip under the wheels can be the difference between rolling out safely or rolling over entirely.

Now, I take nothing for granted. If the wind favors the other direction, I take the extra time to circle. If the grass is slick, I adjust my touchdown point and speed with discipline. And if I'm too fast, I go around - no questions asked.

Because in ag flying, it's not just the wires or the weight or the load that gets you. Sometimes, it's the wet grass you didn't respect.

NOTES:

BOUNCE, BANK, AND BREAK
AIR TRACTOR AT-502B

It had been a long day of ag work. One of those where the heat seems to blur the line between land and sky.

I was flying the AT-502B, hauling fertilizer in and out of a turf strip near Wyoming, Illinois. The wind was light, conditions were good, and I was lined up for what I expected to be a normal landing.

Coming in, everything felt steady. I flared just as I normally would, but the touchdown didn't go as planned. The airplane bounced. Not a little skip - this one was heavy. My instincts kicked in, and I powered up for a go-around.

The engine responded, spooling up quickly. I pulled back and felt the aircraft lift again. But at the top of that bounce, something shifted. The right wing dipped - sharply. I pushed in left aileron to correct, and for a moment, I thought I had it.

I didn't.

Before I could level off, I heard the sickening crunch as the right wingtip caught the top of the corn. The contact yanked the airplane to the right, and we spun. I had no control. The next thing I knew, we slammed down into the cornfield.

The force of impact tore into the aft fuselage and twisted the empennage. The airplane was a wreck. I sat in the cockpit, stunned but unhurt. Alive, but with a harsh lesson burned into my memory.

There hadn't been any mechanical failure. The controls were responsive. The engine performed exactly as it should. The issue had been mine - a misjudged flare, followed by a bounced landing, followed by a rushed recovery that ended with a wing in the corn.

Lessons Learned:

A go-around is a second chance - if you don't waste it.

Bounced landings are part of the job in ag flying. They happen. But how you respond makes all the difference. I should have stabilized the aircraft first before committing to climb out. Instead, I rushed the sequence, reacting too quickly and without full control. That moment of indecision, that failure to correct the dip before the wing caught the corn, was all it took.

Now, every time I bounce - even a little - I ask myself: is this recoverable, or is this a go-around? And if it's the latter, I make sure the aircraft is stable before climbing out.

Flying low and heavy is demanding. It rewards precision - and punishes rushed recovery. When the landing doesn't go to plan, don't try to save face. Save the airplane instead. Go around, but do it right.

NOTES:

INTO THE CORN
WEATHERLY 620A

The light was just beginning to stretch across the Minnesota fields as I lined up for another pass. The Weatherly 620A was loaded and running smooth - just the way I liked it. It was my second job of the morning, and the cornfield ahead shimmered with that golden glow that only comes in early summer.

I descended to working height, ready to begin the next application run. Everything felt right. No wind to fight, no obstacles in the way. I had sprayed this kind of terrain dozens of times. I eased the aircraft down into the pass.

But as I leveled out, something didn't feel right.

I'd misjudged the height of the corn. Maybe it had grown faster than I expected. Maybe I'd been too focused on the swath width or keeping the line straight. Either way, I was lower than I thought. The airplane began to settle just slightly, but that was enough.

The landing gear clipped the top of the corn.

Immediately, the aircraft veered left.

I fought for control, but we were already too low, too committed. In seconds, the left drift carried us right into the field. The main gear dragged through the stalks, and then came the impact.

It was fast and violent. The engine broke free and tore away. The plane spun hard left, ground looping before finally coming to rest upright in the field.

I sat there, heart pounding, staring out at the broken corn and wreckage.

Miraculously, I was unhurt. Shaken, but walking. The airplane hadn't fared as well - both wings were crumpled, the tail section torn and twisted. The engine lay some distance away, a brutal reminder of how unforgiving this job can be.

There was no mechanical failure. No warning light. No weather issue. Just a few inches of miscalculation. A little too low, a little too much trust in routine.

Lessons Learned:

Never underestimate the crop.

Corn grows fast. And when you're flying low, even a few extra inches of unexpected height can make the difference between a clean pass and a crash. I had descended like I always did - based on instinct, routine, and repetition. But nature doesn't fly by checklists. Crops don't wait for you to catch up.

Now, every time I descend into a new field, I take that extra moment. I scan, I measure, and I second-guess. Because the field may look familiar, but the reality on the ground changes every day.

Flying close to the ground gives you no margin. And when the margin disappears, so does control.

NOTES:

THE WINDMILL WAKE-UP
AYRES S2R-T65

It was early - just after sunrise - when I climbed into the seat of the Ayres S2R-T65, ready for another day of spraying. The field outside Wyman looked calm, the sky clear, and the winds light. I'd already circled the area once, checking for obstacles. At the south end of the field stood a windmill - tall, unmoving, a silhouette I made note of but didn't give much thought to.

This wasn't my first time flying around towers, poles, or antennas. I knew how to manage obstacles, and I had every intention of clearing the windmill with room to spare. With my route mapped out, I lined up and began my spray pass.

As I approached the southern edge of the field, the windmill came back into view. I pulled back to begin my climb over it - timing it the way I always had. But that's when I felt it.

The airplane sagged, as if it had been caught in an invisible hand.

I hit what I can only describe as a pocket of dead air - no lift, no forward momentum, just a sudden, sickening drop. In that critical second, I tried to climb harder, but I could feel the resistance. The margin I thought I had disappeared.

The airplane struck the windmill.

It wasn't a catastrophic impact, but I felt the jolt run through the frame. I kept the wings level and turned back toward the airstrip, hoping the aircraft would hold together long enough to get me home.

The approach back in was smooth, deceptively so. I landed gently, easing the airplane onto the runway, but the moment the wheels touched down, the left main landing gear collapsed. The airplane veered sharply off the side and came to rest just beyond the runway edge. I shut everything down and climbed out.

The fuselage was damaged. The gear was gone. But I was unharmed.

I stood there staring at the windmill in the distance and thought about how close I'd come to a very different ending. There had been no mechanical issue, no fault in the aircraft. The engine had performed normally. The airplane had done its job. I hadn't done mine.

Lessons Learned:

You can plan for obstacles - but the air doesn't always play along.

I knew the windmill was there. I planned to clear it. But I relied on assumptions - the kind that build up when you've flown hundreds of hours and made dozens of passes over similar terrain. What I didn't account for was the possibility that the air wouldn't be there when I needed it.

Flying low in variable conditions means respecting not just the obstacle - but the wind, the temperature, and the unpredictable pockets that can shift your margin from "enough" to "none."

Now, I climb earlier. I give myself more space. I treat every spray pass like it's the first. Because out here, near the ground, we fly in a game where inches matter. And when you lose lift, you lose options.

NOTES:

CAUGHT IN THE CORN
AIR TRACTOR AT-802A

It was early morning, the light still soft over the Iowa cornfields as I lined up for another pass. The Air Tractor AT-802A was relatively new, only forty hours on the frame, but it already felt like home. I'd been flying ag ops for years, and that day's mission was straightforward - spray a series of cornfields near Newell, just a short hop from where I'd taken off.

Conditions were perfect. Clear skies, light winds, and visibility for miles. I'd already done a few passes, and things were running smoothly. The rhythm of the work settles in fast - drop down, make the run, pull up, turn, repeat. On one of my final passes, I approached a section bordered by tall trees. I climbed to clear them, just enough to make the turn and come back for the next run.

As I descended back toward the corn, I misread the sink.

The airplane began settling more quickly than I expected. I corrected, tried to level out, but it was too late. The landing gear and belly of the aircraft clipped the top of the corn.

The effect was immediate.

The drag yanked the aircraft deeper into the crop. I tried to climb, but the momentum was gone.

The left wing caught the ground, and the airplane snapped around, rotating a full 180 degrees. The bounce came next - brutal and disorienting. Then a second impact, this time on the right wing and engine. Finally, we came to rest upright in the middle of the field.

I sat there, hands still on the controls, heart racing. Somehow, I was uninjured. Not even a scratch. I unbuckled and climbed out, walking around the aircraft as the adrenaline started to fade.

The damage was sobering. Both wings were crumpled, the engine mounts twisted, and the empennage - battered and bent. It was a miracle the airframe had absorbed the blows without flipping or burning.

There hadn't been any mechanical issues. No warning lights, no strange sounds. Just a misjudgment on descent. A few feet too low. A few seconds too late.

Lessons Learned:

Altitude isn't just a number - it's a decision buffer.

Flying low is part of the job. Every ag pilot knows that. But when you're operating feet above the surface, there's no room for error - no time to recover from a miscalculation. I underestimated the sink rate coming out of that climb. I thought I had more time to settle. I was wrong.

Now, I climb a little earlier and descend a little slower. I build in margins - even small ones - because once you're in the crop, the airplane isn't flying anymore.

In ag aviation, it's not the big mistakes that get you - it's the inches. And on that morning, those inches nearly took the whole aircraft.

NOTES:

SKID TO A STOP
AIR TRACTOR AT-602

The morning started with a light rain - just enough to soak the turf and soften the ground. By the time I was ready to launch for my application run, the weather had cleared, the sky had opened up, and it felt like a go. I was flying out of a familiar strip near Murray, Nebraska - grass, 2,500 feet long, and usually solid underfoot. But that morning, I could feel the softness under the tires during takeoff. Still, I got airborne without issue and headed off to lay fertilizer over a nearby cornfield.

The run itself was smooth - nothing unusual. No wind shear, no gusts, no traffic. I finished the application, turned back toward the strip, and set up for landing.

I touched down as usual, tailwheel and mains settling into the wet turf. As I applied brakes to control the rollout, the airplane responded with something I didn't expect - it started to slide. The right side gave out, the airplane yawed, and before I could fully react, I was veering toward the edge of the strip.

That's when the wing caught the corn.

The right wingtip struck stalks just off the field, and the airplane jerked sharply. In the next second, we ground-looped.

The left wing dug in, the momentum carried us over, and the airplane rolled onto its side.

I sat there, harness tight, dust swirling outside the cracked canopy. I was okay - rattled, but uninjured. The aircraft, though, was a mess. Substantial damage to the fuselage, left wing, and aileron. It wasn't flying again anytime soon.

Post-accident inspections confirmed what I already suspected: no mechanical failure. The brakes had functioned. The engine and airframe were solid. The airplane had done exactly what it was told to do - it just couldn't overcome physics on a wet grass runway.

Lessons Learned:

Grass strips demand respect - especially after rain.

What looks like a routine surface can turn slick in an instant, even with a light rain. A soft turf runway removes your margin. Braking effectiveness drops, directional control becomes a challenge, and all it takes is one unexpected slide to ruin a perfectly good landing.

But more than that, I learned the danger of treating "routine" as safe by default. I knew the strip. I knew the airplane. But I underestimated the conditions.

Now, when I see wet turf, I rethink my braking, my rollout, my touchdown point - and most importantly, my expectations.

Because on that day, it wasn't a gust, a gust lock, or a gear failure that flipped my airplane. It was a slick patch of earth and a split-second slide that changed the outcome. From now on, I treat the ground like it's part of the weather. Because sometimes, the runway is what's waiting to bite.

NOTES:

PART 3 - AIRWORTHINESS & MAINTENANCE

"If you're faced with a forced landing, fly the thing as far into the crash as possible."
    –Bob Hoover

A FUEL GAUGE LIE
UNSPECIFIED AGRICULTURAL AIRCRAFT

I had flown this aircraft for five years and never had reason to doubt it. The fuel gauge had always been spot-on. When it said empty, it was empty. When it said 35 gallons, it meant 35 gallons. At least, I thought it did.

That August morning in Minnesota, I prepared for a typical 45-minute crop-spraying operation. I visually checked the fuel and cross-referenced it with the gauge, which showed 35 gallons. Everything looked good. I launched and worked the fields for about 35 or 40 minutes before heading home.

As I began my climb to ferry altitude, I noticed the fuel gauge had dropped to 20 gallons. Still no concern - until the low fuel pressure light flicked on at just 300 feet AGL. Seconds later, the engine quit.

With no time to second-guess, I turned into the wind and aimed for the best available landing spot. The terrain was rough, and when I touched down, the left landing gear gave out. It tore clean off. The aircraft skidded to a halt, shaken but intact. I got out, unharmed but rattled.

Back on the ground, we drained the fuel tanks completely. Bone dry. Yet the gauge still read 5 gallons.

Lessons Learned:

For five years, the fuel gauge had never let me down - until it did.

Midway through a routine spray flight, the low fuel light blinked on. Seconds later, the engine quit. I made a rough emergency landing and walked away unhurt, but the gear was torn off.

We drained the tanks - empty.

The gauge? Still read five gallons.

That day, I learned the hard truth that even the most reliable instruments can fail without warning.

Don't let familiarity breed complacency. Track time. Cross-check. Verify. Because in ag flying, trust alone won't keep you airborne. Only fuel will.

NOTES:

CONTAMINATED AND COMMITTED
CESSNA 188

While flying a routine agricultural mission, I experienced an engine failure after switching fuel from the wing tanks to the hopper.

At the time, I had 52 gallons of usable fuel in the wings and 40 gallons in the hopper. The engine failed shortly after the switch, and I was forced to perform an emergency landing on a highway at the 21-mile marker. Fortunately, there were no injuries, and the aircraft sustained no significant damage.

The cause of the engine failure was traced back to contaminated fuel. During my preflight inspection, I noticed the fuel had a slight yellowish tint, but I attributed it to the new hopper fuel transfer pump and fuel lines. However, it turned out that the contamination in the hopper fuel was the culprit, which led to the engine failure.

After the landing, A&P mechanics drained the hopper fuel and bypassed all associated plumbing with the hopper fuel lines and transfer pump. The engine was then run-up and checked, and everything was found to be in working order. With FAA approval, I flew the aircraft to a nearby airport after the highway was cleared.

Lessons Learned:

This incident highlights the importance of scrutinizing fuel quality, especially when new components such as a hopper fuel transfer pump are installed. While I initially overlooked the yellow tint in the fuel, the contamination proved to be a critical issue that led to engine failure.

Additionally, this event underscores the necessity of having a comprehensive understanding of the aircraft's fuel system, especially during agricultural operations where fuel switches are routine. A thorough preflight inspection should always include close attention to any anomalies in fuel appearance or other components, regardless of recent maintenance or modifications.

The safe emergency landing on a highway demonstrates the importance of maintaining situational awareness and preparation for unexpected events. In this case, despite the engine failure, I was able to land safely with no injuries, showing the value of staying calm under pressure and executing an emergency procedure efficiently.

Finally, this event reinforced the importance of collaboration with maintenance personnel and FAA approval to ensure that the aircraft is properly restored to service. The quick recovery after the incident was possible because of the team's prompt response and adherence to proper procedures.

NOTES:

CORROSION IN THE SYSTEM
PIPER PA-25-235

The sun had just come up over the fields near Colquitt. The air was already warm, thick with Georgia humidity and the scent of soil. I climbed into the cockpit of my Piper PA-25-235 - an old but faithful Pawnee I'd flown many times before. It was built in '69 and still flying strong. During run-up, everything checked out. Oil and fuel levels were good. No water in the fuel samples. The engine ran smooth.

It was shaping up to be a routine Part 137 spray mission. No clouds, barely any wind, no radio chatter - just the sound of the engine and the rhythm of low-level flight. I launched from the private strip just after sunrise, and the takeoff was clean.

I climbed to about 200 or 300 feet above the ground, scanning ahead for the first pass. Everything felt normal - until it didn't.

I felt it before I heard it: a slight miss. Subtle, but enough to tighten my grip. I looked over my gauges. Throttle forward. Mixture full rich. Prop set. RPMs had dropped - only about 50 revolutions, but it was the wrong direction. I knew something was coming.

And then it came - silence.

The engine quit cold.

No sputter, no stumble. Just quiet. The aircraft had become a glider. I had seconds to act. I looked down - nothing but swamp and dense trees. No clear fields. No options.

I pointed the nose toward a small break in the woods. It wasn't a landing spot - it was the least worst place to aim. I held the glide, flared as best I could, and braced.

The impact was violent. Both wings folded. The fuselage groaned under the force of the landing. We splashed down in shallow swamp water, surrounded by thick underbrush and tall trees.

But I was alive.

Bruised, shaken, but conscious. I unstrapped, pushed the door open, and climbed out. Minutes later, help arrived. They had to cut their way in to reach me.

The aircraft was wrecked. Both wings crushed. The fuselage bent. But there was no fire, and I had walked away.

The investigation came soon after. Everything we could check, we did. The engine showed no signs of failure. The cylinders had compression. The mags worked. The plugs sparked. The crank rotated freely. There was nothing obvious.

Then we looked at the fuel system.

From the outside, the fuel shutoff valve looked fine. But once we opened it, the problem revealed itself. Inside, the valve stem had separated from the plastic internal gate. The stem still turned, but the gate that controlled fuel flow didn't move with it.

Two tiny rivets were the problem.

One had snapped from stress. The other had failed from corrosion - inter-granular, hidden deep within the metal. The valve had shifted slightly out of place, partially blocking the fuel flow. And when the engine demanded full power, it didn't get it.

It was a silent failure. Invisible. Lethal.

The report said it all: fuel starvation caused by an internal valve failure. The engine didn't explode. It didn't quit from misuse. It just stopped - because corrosion had eaten away at a part no one thought to inspect.

Lessons Learned:

This wasn't about flying mistakes. It was about what time does to machines - and how quietly it can kill.

Aircraft can last decades, but time hides its wear. You can't always see corrosion, especially in closed systems. In this case, two tiny rivets inside a sealed valve brought a perfectly good airplane out of the sky.

So what do I take from this?

- Replace aging fuel components before they fail - not after.
- Don't trust five-decade-old hardware in critical systems.
- If you fly older aircraft, assume corrosion exists where you can't see.
- Add the fuel shutoff valve to your regular maintenance inspections.
- Just because something worked yesterday doesn't mean it will tomorrow.

I survived because I flew the airplane all the way into the crash. But I never had the chance to prevent it. That opportunity was lost to a part I couldn't see, failing slowly with every humid morning, every vibration, every drop of time.

Don't let corrosion write the end of your story. Go looking for it - before it finds you.

NOTES:

DEAD STICK OVER THE CORN
ROBINSON R44 II

It was just past dinner time in mid-July. The sky was clear, the wind light, and I was flying low - about fifteen feet over a sea of corn near Pocahontas, Illinois. I was in my R44, making an application pass at about 60 miles per hour. It was the kind of evening that makes you forget things can go wrong.

Until they do.

Without warning, the engine started to cough - just a faint sputter at first, like it was clearing its throat. Then came the silence.

Total power loss.

I instinctively dropped the nose to maintain rotor RPM, flared slightly, and tried to spot a landing area. But I was already over tall, thick corn - ten-foot stalks waving gently in the breeze, offering no clear spot to set down. There was no time. The descent was happening whether I liked it or not.

The skids were the first to go. They caught in the stalks and tore off. The tail boom sheared in the impact. The machine bounced, lifted briefly into the air like a dying breath, skidded forward about thirty feet, and rolled hard to the left.

When it stopped, I was lying on my side inside the fuselage. Still strapped in. Still breathing. A little dazed, but alive.

The wreckage told the story - the tail boom snapped, fuselage crumpled, gear gone. But the big question lingered: why did the engine quit?

Post-crash, we dug into everything. Compression checks, spark plug inspections, fuel lines, magnetos - every square inch of that Lycoming engine. Nothing obvious. No signs of failure. No smoking gun. Ten gallons of fuel on board. Nothing in the teardown suggested a mechanical cause.

And yet, the engine quit.

Lessons Learned:

Sometimes in aviation, the most dangerous words are "it shouldn't have happened."

In our world, we're used to cause and effect - fix the leak, swap the wire, clear the code. But when the engine quits and nothing shows up in the autopsy, you're left with the truth that matters most: you better be ready to fly the aircraft, not just ride in it.

The engine may have failed for reasons unknown. But what saved me was muscle memory. Training. And the decision I made before every flight: to expect the unexpected.

In low-level flying, you're always one heartbeat away from becoming a glider. So fly every pass like it's the one where the noise might stop.

Because someday, it might.

NOTES:

DUCKS IN THE LINE OF FIRE
AERO COMMANDER S2R

Summer flying in South Dakota carries its own rhythm - wide skies, long fields, and the steady hum of a workhorse aircraft tracing low-level lines across the land. That morning, I climbed into the cockpit of my trusted Aero Commander S2R, ready for another spray run over pastureland. With over 5,000 hours in command, this sort of operation was second nature. Low altitude, precise application, and constant vigilance. Just another day in the air.

The aircraft, though old, had been through a thorough inspection a few weeks prior. Everything checked out. The weather was as good as it gets: light wind, smooth air, clear visibility. There was nothing in the forecast, or on the horizon, that gave me any concern. These were the kind of days we hope for in this line of work.

I got airborne mid-afternoon, and the job was moving along smoothly. I'd completed several passes, each one uneventful. Just me, the hum of the engine, and the field unfolding beneath my wings.

As I climbed out after a run, maintaining about 30 feet above the ground, the unexpected happened.

A sudden burst of movement caught my eye. A flock of mallard ducks exploded into view directly in front of me.

They came from low and to the side, fast and heavy. I had no time to react. Before I could adjust or bank, I flew straight through them.

The impact was immediate. I heard the engine change pitch, then sputter. And then silence. My power plant had just gone quiet. In that moment, the aircraft became a glider - and not a particularly graceful one. At roughly 80 feet, I didn't have many choices. I didn't have time to dump the chemical load, nor altitude to circle back. I scanned ahead, picked the most forgiving patch of ground I could see, and committed.

The landing came quickly. I leveled the wings and aimed straight. The wheels touched down, but instead of firm pasture, I found soft, wet soil. The main gear sank deep - like hitting a sponge. The nose dipped, momentum carried the aircraft forward, and before I knew it, we were upside down. The cockpit filled with the scent of earth and fuel. I hung there, strapped in, disoriented but alive.

My first instinct was to get out. The right-side door was blocked, packed with mud. I looked left - broken glass. I kicked out what remained of the window, braced myself, and crawled through. I came out dirty, bruised, but walking.

Emergency crews arrived quickly. I was treated for minor injuries and released that same day. The wreckage was extensive - twisted fuselage, cracked engine mount, structural damage to the right wing. But I'd made it out alive. That's all that mattered.

Later analysis confirmed what I already suspected. The engine had ingested multiple birds. Mallards - both male and female. DNA tests were conclusive. There were no mechanical issues, no missed inspections, no overlooked faults. The aircraft had been sound. The ducks, unfortunately, had been in the wrong place at the worst possible time.

The final assessment was straightforward: engine failure due to bird ingestion, followed by a forced landing that ended in a nose-over. No pilot error. No fault assigned. Just an unavoidable encounter between machine and nature.

Flying ag ops under Part 137 means flying in the margins - those tight vertical bands just above the ground where reaction time is measured in seconds, not minutes. There's no tower to warn you, no altitude buffer to rely on. Just instinct, skill, and hope that nothing crosses your path unannounced.

Lessons Learned:

Even on perfect days, nature doesn't play by the rules. One moment I was flying a routine spray pass; the next, a flock of ducks turned my aircraft into a glider.

The engine failed instantly from bird ingestion. There was no time to dump the load or find better ground. The forced landing ended in a nose-over, leaving the aircraft wrecked. But I walked away.

The lesson? Birdstrikes aren't rare flukes. They're real threats in low-level ops.

Stay alert, know your outs, and never let smooth conditions lull you into complacency. Because even a duck, in the right place at the wrong time, can bring you down.

NOTES:

ENGINE OUT, OPTIONS THIN
SCHWEIZER G-164B AG CAT TURBINE

During a ferry flight at 500 feet AGL, I was operating a Garrett Turbine-powered Schweizer G-164B Ag Cat under FAA Part 137 operations. Approximately 45 minutes into my one-hour flight, I noticed a significant loss of power. The torque gauge dropped from 42 to 28 almost immediately, with 28 being the lower end of the green arc for normal power settings. The torque remained at 28 for about 20 seconds before all power was lost. Throughout this drop in torque, the engine RPM remained steady at 100%.

Recognizing the critical situation, I declared an emergency and began looking for a suitable landing site. I identified a road with no power lines or vehicles and made an emergency landing. Unfortunately, while rolling into the ditch, the wings made contact with the ground, causing slight bending at the ends, and the propeller blades were also bent. Despite the damage to the aircraft, the airframe itself was unaffected, and there were no chemicals onboard.

I was able to exit the aircraft safely, and emergency services (police, ambulance, and fire trucks) arrived immediately on scene.

Fortunately, there were no injuries to myself or anyone else, and no property damage occurred. The FAA was notified immediately after the incident.

Lessons Learned:

When ferrying an ag aircraft, especially one operating under turbine power, staying attuned to even subtle changes in performance is critical.

In my case, a sudden drop in torque with steady RPM signaled deeper engine trouble that quickly escalated into total power loss. I had to act fast - spotting a safe landing area and executing an emergency landing with no injury and minimal damage.

This reinforced how essential it is to always have an emergency plan in mind, even on routine legs. Thorough pre-flight checks matter, but so does maintaining a calm, decisive mindset when things go wrong.

You only get one chance to get it right.

<u>NOTES:</u>

FIFTEEN FEET FROM FAILURE
SCHWEIZER GRUMMAN G-164B AG CAT

It was my fifth spray flight of the day. The kind of rhythm you get into when the job is flowing, the sun's high, and the winds are behaving. I taxied the Schweizer G-164B into position at the grass strip outside Knox, loaded and ready to go. I'd flown this airplane for five years - reliable, no surprises, part of my body at this point.

Takeoff was standard. I rolled down the strip, watching the end get closer as I built up speed. Around fifteen feet above the ground, I expected the usual smooth climb. But something was off. The plane wasn't climbing.

I nudged the throttle forward. Nothing. No power surge, no response. Just that awful feeling of dragging through the air, nose up, going nowhere.

Ahead of me, corn.

Still in a nose-high attitude, the right wing dipped and caught the field. That was it. The drag yanked the aircraft sideways, and we hit the ground with brutal force - ripping through stalks, dirt, and metal.

When we stopped, I was sitting upright in a broken machine, heart pounding. But I was unhurt. Not a scratch.

The aircraft wasn't so lucky. Both wings were battered, the fuselage twisted, the propeller trashed. The engine bay was caked in earth.

At first, I thought the engine had failed. It had to be a mechanical issue, right? The G-164B had been inspected just weeks before. A new prop governor had even been installed. But the teardown told a different story.

No sign of failure, or broken components. Just the kind of mess that comes after a hard impact with the earth. Some odd contaminants showed up in the fuel control unit - RTV silicone sealant where it didn't belong, and debris, but nothing that explained the loss of power. The engine had been running when we hit.

That was maybe the hardest part. There was no smoking gun. No clear failure to point at and say, "That's what got me." Just silence and uncertainty.

Lessons Learned:

Not every loss of power comes with a warning light.

I still don't know what caused the partial loss. Maybe something in the fuel line. Maybe a sticky component deep in the FCU. But what I do know is this: at 15 feet, there's no time for diagnosis. You only get one shot to fly - or fall.

This experience changed my takeoff mindset. I now treat every departure like it might be the one where things go quiet. I stay alert for any lag in performance, any hesitation in the climb. And I don't hesitate to abort if something feels wrong.

Just because you've flown an airplane for years without a problem doesn't mean the next flight won't be the one that surprises you. Machines hide their problems well - right up to the moment they don't.

Respect every takeoff. Know your emergency options. And never assume you've got more altitude than you really do.

NOTES:

FIRE IN THE FIELD
AIR TRACTOR AT-301

I had just bought the airplane the day before. A new addition to the fleet, freshly ferried home and put straight to work in the skies over Tennessee. I'd already run a few flights with no issues. Nothing jumped out during the initial inspection, and aside from a little backfiring during a magneto check - which we cleared up with some cleaning and plug replacements - everything seemed fine.

It was a warm May day, and I was mid-run on my fourth load of the afternoon. The kind of flying I've done for years - low, focused, methodical. But that's when things changed.

I felt it before I saw it.

A sudden wave of heat against my legs, like someone opened a furnace under the floorboards. Then a flash - just the briefest glimpse - out of the corner of my eye. A lick of flame on the left side of the cowling. I rocked the wings to get a better look, hoping I'd imagined it. I hadn't.

There was a fire.

I had no time to think. Just react.

I reduced power and threw the flaps full down to slow her up. Scanned for a place to land. Nothing obvious.

I spotted some truck tracks across a rough patch of ground, maybe an old turnrow. It wasn't ideal, but it was something.

I made a tight circle to lose speed, slipped in from the right, and leveled out just before touchdown. The landing was fast, too fast, but I was more worried about getting on the ground than greasing it in. We hit hard, rolled maybe 100 feet, and then caught a rut. The gear buckled and we nosed over. Metal shrieked. Dirt flew.

The fire didn't stop.

By the time I scrambled out, flames were working their way aft from the cowling. I ran. There was nothing else to do.

Later, they tore the wreckage apart. What they found was ugly.

The fire had started just behind the engine, low on the left side near the exhaust manifold. One of the clamps securing the exhaust segments was missing. That section had been leaking hot gases, probably for a while. The backfiring I'd noticed earlier? Most likely the first sign.

Those leaking exhaust gases had been blowing directly into the cowling. Over time, heat had degraded the oil hoses. One of them - the oil tank outlet hose - failed under fire. Once it did, it turned a small burn into an oil-fed inferno.

There were more problems. Other clamps were loose. The exhaust manifold showed burn-throughs, leaks, even holes. All of it suggested long-term neglect. And yet the airplane had just passed an annual inspection less than ten days earlier.

According to the manual, the exhaust system should be checked every 50 hours for signs of burning or cracks. FAA rules say the same - inspect the stacks, check for defects, check for loose fittings.

But no one had caught it. Not the seller, the mechanic, or me.

Lessons Learned:

Just because it starts, doesn't mean it's airworthy.

This aircraft flew fine... until it didn't.

The fire didn't come from a fuel line rupture, or pilot error, or electrical fault. It came from an overlooked exhaust leak and a worn hose. And all of it was hiding in plain sight.

Buying a new-to-you airplane, especially a radial engine workhorse like the AT-301, comes with risk. You can't assume that because it flies, it's been cared for. You can't trust the paperwork alone. And you certainly can't let a clean annual lull you into complacency.

From now on, I treat every new airplane like a suspect.

I crawl under it. I get my hands dirty, and I look at the things no one talks about - the clamps, the hoses, the ugly stuff under the cowling.

Because when fire erupts at a hundred feet, the checklist doesn't matter. You're out of time. And out of options.

NOTES:

FUEL TO NOWHERE
AIR TRACTOR AT-401

The day had started like so many others. I was halfway through a routine application job, cruising low and tight over a field just outside Carroll. The Air Tractor AT-401 beneath me was steady, humming along as it always had. Nothing out of the ordinary. Blue skies, calm winds, a clean pass - ag flying at its most straightforward.

Then, without warning, the hum dropped into silence.

Just as I rolled out of a turn to line up for the next spray pass, the engine lost power. It wasn't a sputter. It wasn't gradual. It was as if someone had pulled the plug.

I had no time to troubleshoot. I dropped the nose, scanned for a spot, and committed to a forced landing. The nearest open ground was rough and tight, but it was all I had.

I brought the aircraft down as smooth as I could. The touchdown was hard. Both wings caught the terrain. The structure groaned under the impact, but the fuselage held. I came to a stop in a mess of dirt and metal, shaken but uninjured.

I climbed out, my boots crunching into the earth, and looked at what was left of the aircraft.

The wings were bent, crumpled from the impact. The engine looked untouched, but something had clearly failed.

When investigators arrived, they began the slow work of piecing it together. The engine itself - nothing wrong. The fuel pump? Fully functional. Fuel control? Passed all tests. That's when they turned their attention to the airframe fuel boost pump.

On the ground, the test was simple: activate the pump and check for fuel flow. There was none.

A few lines were loosened and checked - fuel was in the header tank. But with the pump running, the fuel wasn't moving. There was no flow, no pressure, and no bypass function. Nothing. Just silence where there should've been motion.

The pump was pulled out and sent to its manufacturer for bench testing. What they found was telling. The shaft that connected the electric motor to the pump mechanism was so worn it didn't even engage. The seal had failed too - fuel was leaking. The motor spun, but nothing turned. It hadn't worked at all.

Turns out the pump had last been overhauled eight years earlier - outside the recommended 10-year overhaul cycle. There was no hour-based recommendation - just a time limit. One that had expired.

It wasn't a flashy failure. It didn't explode. It didn't smoke. It simply wore down - until one day it couldn't do its job. And when I needed it most, there was no fuel getting to the engine.

Lessons Learned:

Some failures whisper before they scream.

This wasn't about pilot error, or a blown engine, or even a fuel miscalculation. It was about a simple, overlooked part - one that seemed fine until the moment it wasn't.

We often focus on the big pieces: turbines, props, wings. But in aviation, the smallest component can bring everything down. That boost pump? It was just a support role. But without it, the rest of the machine was useless.

Now, I don't just check the usual suspects. I look at the maintenance logs with a sharper eye. I ask about the small parts - the ones hiding deep in the systems that never get a second thought until they quit.

Because sometimes, it's not a loud bang or a flashing light. It's a worn shaft, a slow leak, and a silent motor spinning nothing but air.

NOTES:

INVISIBLE ICING
AIR TRACTOR AT-301

The morning was clear and cool - perfect for a spray run. I took off from a private strip near Circle, Montana, in my Air Tractor AT-301. It's an aircraft I knew well. I'd flown it for decades, logged thousands of hours in it, and this job was as routine as they come.

The radial engine sounded strong, the visibility was excellent, and I expected to be up and back in minutes.

The mission itself went smoothly. I completed the spray run without issue, and then turned back toward the strip. At about 100 feet above the ground, just a few minutes from home, I felt the engine begin to run rough. Then it lost power completely.

There wasn't time to think. No altitude. No margin. I immediately looked for a landing spot and saw a wheat field just ahead. I aimed for it and began the forced landing. I knew the chances were slim. At this height, you don't float - you fall.

I came up about 75 feet short.

The aircraft hit hard. The fuselage absorbed the impact with a jolt that rattled my bones. But the airplane didn't catch fire, and when everything stopped moving, I realized I was unhurt. Shaken - but walking away.

Later, as I reviewed what happened, it hit me: I never once considered carburetor ice. I didn't even think to apply carb heat. The weather had seemed "fine." But I should've known better.

When the investigation wrapped up, they found no mechanical failures. Nothing broken. No fuel blockage. No linkage issues. The engine was intact. The systems were working. It wasn't the machine - it was the conditions. And the omission.

The nearby weather station showed a temperature of 14°C and a dew point of 12°C. Not dramatic - but just right for serious carburetor icing, especially at cruise power. The kind that sneaks up on you.

My engine was one of those big carbureted radials - known to be vulnerable. The protection is simple: carb heat. But it only works if you use it before trouble starts.

And I hadn't.

The conclusion was straightforward: loss of engine power due to carburetor ice. Contributing to the accident was my failure to use carb heat. It's a hard sentence to read about yourself, especially after so many years in the air.

But they were right.

Lessons Learned:

Carburetor icing doesn't shout. It whispers. And if you don't listen, it takes your engine - quietly, completely.

Here's what I'll never forget:

- Carb ice forms in weather most pilots would call "good." Clear skies don't protect you.
- If temperature and dew point are close, assume the threat is real.
- Carb heat isn't optional. It's the only line of defense.
- Apply it early - especially in low-power cruise or high-humidity conditions.

- Don't wait for symptoms. By the time the engine sputters, it might be too late.

I've flown through a lot of weather, over a lot of terrain, in a lot of aircraft. But this wasn't about experience. It was about routine. Familiarity. Maybe even complacency. I assumed I didn't need the extra step. I was wrong.

The aircraft was damaged. The job wasn't finished. But I walked away.

That morning reminded me that aviation doesn't reward memory, it demands discipline. Every flight, every time.

The procedures you think you've mastered are the ones that can bite when forgotten.

And sometimes, it's the smallest oversight - just one switch not moved - that turns a simple flight into a crash.

NOTES:

LIGHTS OUT IN THE CANYON
BELL UH-1B HUEY

It was late afternoon with dusk creeping in fast when I departed Joseph Plains en route to Clarkston, Washington in a UH-1B Huey helicopter set up for agricultural operations. The flight was uneventful at first. The aircraft's rotating beacon and position lights had all been checked preflight and were working fine. The rotating beacon stayed on, and as the sky dimmed halfway through the trip, I flicked on the position and cockpit lights. From inside the Huey, everything looked good.

What I couldn't tell was that the position lights were no longer operational.

The final leg of the flight took me down the Snake River Canyon, around 7–10 miles upriver from Clarkston. Visibility was starting to go: overcast skies, haze, patches of light rain. With the shadows lengthening, I reached for my handheld radio to contact Lewiston Tower. But on transmit, the battery fizzled out. No confirmation, no two-way - just silence.

Down in the canyon, below the rim and hidden from the tower's view, I had a decision to make.

I could try to land short of the Class D airspace, or continue on to my known landing zone in Clarkston. I weighed the risks - unfamiliar landing zones in dim light and the possibility of wires or terrain hazards I couldn't see versus a known, lit location near town. I chose the known.

I slipped under the Class D shelf, cruising about 400 feet below the airport elevation, and continued down the canyon. I didn't want to use the landing light, worried it might spook motorists on the new road nearby, but the fading daylight was just enough to guide me in. I landed safely in a commercial storage yard along the river, unaware of the issues that had quietly stacked up during the flight.

During cooldown, I leaned out of the aircraft and saw it: no position lights. No rotating beacon. Completely dark.

The next day we did a full inspection. One position light had shorted and popped the breaker, knocking them all out. The beacon had burned out mid flight. The deck light had moisture intrusion. And the handheld radio? Dead battery. To top it off, the temporary registration paperwork had expired - though we'd filed a second application and contacted FAA Registry to follow up.

Lessons Learned:

This wasn't a case of negligence, but one of compounding small oversights.

The aircraft hadn't flown in over two months and had been exposed to the elements. Lights that worked fine during the preflight failed en route. My handheld radio, the only link to ATC, was undercharged. And though FAR 137.47 allows agricultural ops without position lights, the intention was never to operate without them at nightfall.

The truth is: even with experience, a mission can slide into gray areas when the pressure is on and the light is fading.

I took the safest route I could in the moment, but hindsight reminds me that preparation - and redundant backups - are what really keep us safe.

Now, we're installing a permanent comms system, improving our maintenance checks, and treating dusk like what it really is: the opening act of night.

NOTES:

OUT OF FUEL
GRUMMAN G-164B

It was supposed to be the final run of the day - just one more load before I could call it and head back in. The Grumman G-164B and I had spent hours in the sky already, treating rice fields around Hoxie. I was tired, sure, but focused. That's how it is with ag flying. You find a rhythm and lock in.

As I taxied out with the last load, I noticed the fuel level wasn't where I wanted it to be. It was low. Not empty, but lower than comfortable. I considered heading back to top off, but I reasoned I could get through the load. Just a quick job. Fifteen, twenty minutes max. I knew the numbers. I'd done it before.

So, I went.

Flying low over the green floodplains, I moved with precision, watching for obstacles, adjusting for wind, keeping the lines clean. I was halfway through dispersing the load when it happened.

The engine coughed once, then went quiet.

Instantly, my heart kicked into overdrive. I pushed the throttle, flipped the switches, but I already knew - it was out of fuel. The sound of silence in the cockpit is deafening when it shouldn't be there. I was running on gravity and glide now.

I scanned the terrain for options, but there were few. The field I was working was flooded - typical for rice - but less than ideal for a forced landing. Still, it was better than the treeline or the levees. I aimed for the smoothest patch I could find and brought the nose up gently as I descended.

The wheels touched down in water and soft mud. I thought for a moment I might roll out cleanly. Then the left main gear dug in, caught the soft bottom, and the whole aircraft flipped.

Everything blurred - the belly of the aircraft was suddenly above me, the harness held tight, and the cockpit filled with water and noise. When it stopped, I was upside-down, disoriented, but alive. I managed to release myself and crawled out, soaked, shaken, but unhurt.

Looking back at the aircraft, it was clear the damage was severe. The right lift strut had buckled, the fuselage twisted, the wings sagging. The airplane was done. And it was my fault.

The investigation confirmed what I already knew: the engine had lost power due to fuel exhaustion. There were no mechanical issues. Nothing had failed except my decision-making. The G-164B had been in good shape. It had recently passed inspection. The engine was strong. But none of that mattered when the tanks ran dry.

Lessons Learned:

Assumptions can be louder than alarms - and far more dangerous.

I knew I was low on fuel. I thought I had enough. I calculated in my head instead of measuring in fact. In this business, we often pride ourselves on instinct, experience, and feel. But feel doesn't move fuel through the lines. Numbers don't lie, and neither do empty tanks.

Now, I don't push it - not for one more load, not for five more minutes. If the fuel is questionable, the answer is simple: land, refuel, reset. Because it's not just about finishing the job - it's about making it back.

Flying ag is unforgiving. There's no altitude cushion, no glide path, and no margin for error when you're skimming treetops or dropping into waterlogged paddies. Every decision counts.

That day, I gambled on my judgment - and I lost. The aircraft paid the price. I didn't. But next time, it could be different.

The tanks don't lie. Check them. Trust them. Respect them.

NOTES:

OUT OF FUEL, OUT OF OPTIONS
AIR TRACTOR AT-602

The afternoon sun was high, and I was partway through another aerial application flight over the Sacramento Valley. The field looked familiar - flat, dry, and easy to navigate. I'd flown this route more times than I could count. It was supposed to be a routine run, quick and efficient. But routine can be dangerous when it dulls your edge.

I was maneuvering at low altitude when the engine quit.

No warning. No stutter. Just silence.

I instinctively went into restart mode - checked everything I could, flipped switches, looked for pressure. Nothing. The prop windmilled uselessly, and the ground rushed up fast. With no time to return to base, I aimed for an open patch of plowed field and prepared for impact.å

The touchdown was hard. The left wing struck first and took most of the damage. Dust exploded into the cockpit as the aircraft slid to a stop. I was okay. Shaken, but not hurt.

As I climbed out and looked at the broken wing, the reality hit me: something had gone seriously wrong. I hadn't heard any unusual engine noise, and everything had felt fine just moments before the power loss. So why had it quit?

Later, when the investigators got to work, the answer was simple - and embarrassing.

The right fuel tank was nearly dry. The left had about 15 gallons left in it. The aircraft had flamed out due to fuel starvation.

The manual is clear: below half tanks, skidding turns can cause fuel to shift from one tank to the other. If one side runs dry and the feed isn't balanced, you lose fuel flow - and the engine dies. That's exactly what happened. In all my years flying, with over 31,000 hours in the logbook, I'd made a rookie mistake.

I hadn't inspected both tanks thoroughly before takeoff. I had assumed. I'd glanced in, done a quick visual, and gone on muscle memory. I thought I had enough fuel for the run, but I didn't calculate the distribution or check carefully enough. And in that moment, all the experience in the world couldn't save me from what I didn't know.

Lessons Learned:

Experience never replaces discipline.

When you've flown for decades, you learn to read your machine, trust your instincts, and operate by feel. But comfort is not a substitute for checklist discipline. I failed to inspect the tanks properly - and I paid for it.

Fuel starvation is preventable. It's not a mechanical failure or a hidden fault. It's a pilot's job to know where the fuel is and how it moves, especially in ag flying where maneuvering is tight and conditions can shift everything in seconds.

That day, I learned that you don't outgrow the basics. You don't fly enough hours to skip the fundamentals. And you never, ever assume when it comes to fuel.

I was lucky. I had open ground and walked away. Next time, the margin might be thinner.

Now, I take the time.

I stick the tanks, I check each side, and I fly like I haven't seen this field before.

The minute you treat a flight like it's routine, you invite surprise.

And some surprises don't let you land.

NOTES:

OVER THE LINE
GRUMMAN G-164B AG CAT

It was May, and by the time I climbed into the cockpit for my 12th flight of the day, the rhythm had set in. Load, spray, return. Repeat. The sun was high, the air thick with heat and humidity. I was pushing hard, but that's how ag flying goes in the busy season.

The load crew was working fast, maybe too fast. They pumped fertilizer into the hopper, but something felt off. I glanced at the weight, did the mental math, and knew - this load was heavy. Too heavy. The ground equipment wasn't reading right, and I could see it in the hopper level. We'd gone over.

I should've stopped everything. But instead, I made a quick decision. I'd release some product during takeoff. Just enough to bring the weight down before I really got airborne.

I lined up on the grass strip, engine humming, tailwheel bouncing gently as I pushed the throttle forward. I let the aircraft roll longer than usual, waiting for that moment when it felt like it wanted to fly. As I hit that spot, I started dumping fertilizer - lightening the load, hoping for lift.

The airplane responded. Sort of.

It staggered into the air, clawing for altitude.

I nursed the stick, coaxing every ounce of lift from the wings. As I turned toward the field, trying to settle into the application run, the aircraft betrayed me. It didn't climb. It didn't hold. It just... sank.

We hit the field - hard. The wings crumpled. The empennage snapped. The engine tore away from the fuselage. Metal bent around me like foil. But somehow, I wasn't hurt.

After the crash, I admitted what I knew from the moment I took off: the airplane had been overloaded. I had rolled the dice and lost. The airframe had no mechanical faults. The engine was working. The airplane was fine until I asked it to do something it couldn't.

Then came the paperwork. There was no Part 137 certificate for the operator. No skill test recorded for me, either. I hadn't met all the requirements for ag operations. And that cut deeper than the crash.

Lessons Learned:

There's a difference between experience and qualification, and between knowing the risk and respecting it.

I gambled with margins, letting a faulty ground operation and a rushed schedule pressure me into compromising safety. I flew an aircraft that was too heavy and relied on a shortcut to fix it mid-roll.

Dumping fertilizer on takeoff isn't a procedure. It's a last-ditch move. And I treated it like a strategy.

We're used to working the edge. But there's a line between operating efficiently and ignoring limits. When you cross it, the airplane doesn't give you a second chance - it gives you a crash.

I now treat every load like a commitment. If it's wrong, I fix it before I fly. And I make sure the paperwork matches the reality. Because when things go bad, the logbook matters almost as much as the stick and rudder.

Sometimes, the only thing between you and the dirt is your own judgment. Make sure it's worth trusting.

NOTES:

POP, SMOKE, AND FENCE
AIR TRACTOR AT-502B

Just after sunrise, I was climbing out from Decatur with a full load. It was a clear July day - perfect for spraying - and everything about the takeoff felt routine. I'd done it many times before. The Air Tractor AT-502B roared up the runway, eager and strong. I had just rotated and begun the shallow climb when it happened.

A sharp *pop*.

Immediately, I noticed white smoke trailing behind the aircraft. The power was gone - or going. My instincts kicked in. I nosed down to hold speed and started looking for a spot to put her down.

I didn't have time for much.

I aimed for a bean field just past the airport boundary. There was no altitude to work with, no margin for error. The engine had gone quiet, and I was dropping fast. I cleared the runway, glided toward the open field, and set her down as smooth as I could manage.

But I couldn't avoid the airport perimeter fence.

We hit it.

Both wings were torn up in the impact. The airplane skidded and stopped abruptly in the soft crop rows.

I shut everything down, climbed out, and stood in the field staring at the wreckage - shaken but physically fine.

Later, the investigation found what I already suspected: catastrophic engine failure. The compressor turbine blades had fractured - every single one of them. Some had cracked and shattered at different lengths, and the downstream power turbine blades were wrecked from the cascading damage. Metallurgy showed clear signs of creep - tiny voids in the material grain structure - and overheating, with signs the turbine blades had been exposed to temperatures beyond engine limits.

It wasn't a sudden defect. It had built up over time.

What I didn't know - and what no manual had screamed loudly enough - was that repeated use near or above the recommended power settings, especially under tough working conditions, was slowly cooking those blades from the inside out. We were right on the edge for too long, and that morning, the edge gave out.

Lessons Learned:

Engines don't break all at once. They surrender, piece by piece.

Ag pilots lean on their equipment hard. Every day is a test of airframe, engine, and nerve. But the truth is, no matter how solid the machine seems, it remembers every second of how you fly it. That turbine had been pushed close to its thermal limits one too many times.

Since that day, I've changed how I look at engine data - especially temperature and torque. I stay within the book. And more importantly, I give the engine space to breathe when I can. It might cost me a few seconds on the climb or a little more time in the day - but it's better than staring at a fence line with dead silence behind the prop.

Creep isn't loud. It doesn't warn you. But when it wins, you fall.

NOTES:

POWER LOSS, CLEAR THINKING
UNSPECIFIED AGRICULTURAL AIRCRAFT

It was late in the day when I launched from MTW on an agricultural chase mission, flying in coordination with Wisconsin state observers. We were en route to a spray area when, just southwest of Kewaunee, I noticed the engine begin to run roughly. Something you never want to hear when you're 2,500 feet up!

I immediately ran through the emergency checklist for engine roughness and began troubleshooting. Pulling back the throttle seemed to help-the engine smoothed out around 1700 RPM. But that didn't inspire much confidence. I wasn't about to risk pushing it, only to suffer a total engine failure over unwelcoming terrain.

One of the spray pilots on the radio suggested something bold but practical: find a road - any of the east-west highways below - and set it down while I still had some power.

Highway 29, just three miles southwest of Kewaunee, looked promising. After carefully watching for traffic, I committed. The landing was smooth, controlled, and uneventful. I taxied off the road and into a nearby farmer's driveway, shutting down with oil pressure still showing healthy.

Post-flight inspection revealed the culprit: a bent pushrod due to a frozen valve. The engine still had five quarts of oil left in the sump.

Lessons Learned:

When the engine ran rough at 2,500 feet, I had a choice: push on and hope-or act early while I still had options. Reducing throttle helped temporarily, but I knew partial power can vanish in seconds. I opted for a safe, controlled landing on a nearby highway.

The diagnosis? A bent pushrod from a frozen valve-proof that failure wasn't far off.

This experience reinforced a simple truth: don't wait for the engine to quit completely before deciding. When power problems begin, act fast, stay calm, and aim for a known surface.

A road under your wheels beats a field under your fuselage - especially when time and altitude are slipping away.

NOTES:

THE BANG AND THE BOUNCE
AIR TRACTOR AT-502B

The day had been long, but smooth. I'd already completed several spray runs, and I was on my sixth load. The Air Tractor AT-502B had been running like a dream all afternoon. Fueled and topped with chemicals, I lifted off from the strip near Tappahannock and worked the fields near Center Cross.

Everything about that pass felt routine. Until it wasn't.

I had just completed the final pass of the load. As I climbed out to reposition, about 350 feet above the trees, I heard a loud bang.

In an instant, everything changed.

The engine lost power. Not a gradual drop - not something I could manage - but an abrupt loss. My instincts kicked in. I lowered the nose immediately and started scanning for somewhere - anywhere - to land. Trees, ditches, and rough terrain closed in fast.

I spotted an open patch between the trees. It wasn't perfect, but it was the best I had. I committed.

Coming in, I knew I had too much speed. I tried to bleed it off, but the airplane floated. The first touchdown was hard, and we bounced.

I held on as we skipped forward and slammed into the treeline. The wings crumpled. The airframe groaned. But when everything stopped, I was still in one piece.

Upright. Alive.

I climbed out, heart racing, staring at the broken machine that had carried me through thousands of hours of flight. The fuselage was buckled. Both wings were crushed in. The right elevator was bent, the vertical stabilizer twisted sideways. But no fire. No fuel leak. Just silence, and questions.

What had failed?

Back at the hangar, they combed through the wreckage. There was plenty of fuel onboard. The control continuity was intact. The engine hadn't exploded or seized. The propeller showed signs of rotation - some abrasion on the blades, even branch impacts that suggested it was spinning when we hit. But there were no clear answers.

The turbine had some rubbing - indicating low power at impact - but no evidence of internal failure. The fuel system was clean. Chip detectors and filters were all normal. One compressor turbine blade had broken off, but it wasn't enough to explain the whole event. Everything pointed to a power loss. But why?

That was the mystery. One that never got solved.

Lessons Learned:

Not all failures come with an explanation.

I'd done everything right - or thought I had. Full fuel. Full checklist. No warning lights. But when that bang came, I had seconds to act.

In ag flying, you plan for the worst and fly with the best. But sometimes, the airplane surprises you - and not in a good way. What saved me that day wasn't luck. It was preparation.

I'd practiced forced landings. I knew what to look for. I gave myself options, even if they weren't great.

Now, I take nothing for granted. Every climb-out is a potential glide. Every pass might be my last one with power. And every engine, no matter how well it's maintained, carries the potential to stop.

Flying is about staying one step ahead of what you can't control. Because sometimes, when the engine quits, the only thing left working is you.

<u>NOTES:</u>

THE ENGINE THAT LET GO
GRUMMAN G-164B AG CAT

It was the sixth takeoff of the day, and the rhythm was smooth. I'd started flying just after sunrise, loading fungicide and fuel, making pass after pass over the fields outside Amana. The weather was clear, the wind light, and the airplane - a rugged old Ag Cat - was running well. I was loaded with 420 gallons of mix and 110 gallons of fuel for this leg. Everything about the departure felt routine.

The airspeed built up steadily. At about 65 mph, I eased back and let the aircraft break ground. We climbed out in ground effect, gathering speed, ticking past 78 mph. Then it happened.

The engine quit.

No warning. No roughness. Just a sudden, sickening silence where there had been a strong, steady roar moments before.

I was maybe a few hundred feet up - nowhere near enough altitude to turn back. I pushed the nose down gently, aiming straight ahead toward the open cornfield off the departure end of the strip. It was the only option I had. The airplane settled fast, like a dropped stone. I pulled back just before impact, trying to arrest the descent, but the corn swallowed the wheels and the nose went down hard.

We flipped.

The airframe groaned and twisted as we rolled over. When we stopped, the world was upside down - literally. I unbuckled and crawled out through shattered glass and crushed metal. The wings were bent, the fuselage was warped, and the propeller blades were curled back like ribbon. But I was alive. I hadn't even been scratched.

Later, the engine was pulled and torn down. It was a Honeywell TPE331, and everything looked fine - at least at first. No damage in the fuel control unit. No governor issues. But when they dug deeper, they found the heart of the problem.

The torsion shaft had failed.

It had broken inside the engine, creating a disconnect in the drive system. Without that connection, torque couldn't transfer. The engine was technically still running, but it couldn't drive the propeller. It wasn't something you could see coming - not from the outside. There had been no signs of wear, no strange readings, no prior symptoms.

The worst kind of failure: the kind you can't predict or prevent.

Lessons Learned:

Not every failure gives you a second chance - but you have to be ready as if it might.

In this case, the airplane failed me - but my decisions saved me. I didn't try to turn back. I didn't chase altitude I didn't have. I aimed straight, kept my head, and managed the crash landing the best I could. And because of that, I walked away.

There are times in this business when you won't get a warning. You won't hear a rattle or see a gauge drop. You'll just lose power - and the rest will come down to training, instinct, and what you choose to do next.

We like to think we're in control. But sometimes, the machine makes the final call.

That's why every takeoff matters. Every loading decision, every scan of the gauges, every mental note of wind and weight. Because someday, you'll need every bit of it.

And when the power goes out, what's left is what you put in before the engine ever turned.

NOTES:

THE FUEL THAT FOAMED
BELL OH-58A

It was a cool morning in May, and I was a few runs into a long day of aerial spraying. The Bell OH-58A had been running smooth all season - solid engine, reliable handling, nothing fancy but tough and dependable. That day was no different. I'd topped off from the usual support truck and lifted off from a strip near Wade, Oklahoma, like I'd done so many times before.

The field was familiar, and I was just finishing a low maneuvering pass when it happened.

Without warning, the engine lost power. It didn't sputter or surge. It just quit cold.

There was no time to troubleshoot. I immediately entered autorotation, searching for any patch of field that might absorb the impact. My hands moved on instinct, muscle memory honed by decades in the cockpit. I flared as best I could, but the hard landing drove the skids deep into the dirt, and the rotor torque twisted us. The helicopter rolled, landing on its side in a shower of dust and debris.

I came out of it with only minor injuries. The aircraft didn't fare so well. Substantial damage - blades, tail, airframe. The kind of damage that ends a machine's career.

The investigation started quickly. We checked the engine and fuel system. What we found stopped me cold: foamy, cream-colored liquid in the fuel lines and filter. That wasn't just fuel - it was fuel contaminated with something else. Water? Oil? Emulsified gunk? No one could say for sure.

I'd refueled that morning from the support truck - the same truck I'd used the day before. At first, we found no visible contamination in the truck tank. But when we looked again days later, there it was: the same creamy sludge we found in the helicopter.

Where did it come from? The truck had been filled from the on-site fuel storage tank, but we couldn't trace the source of the contamination beyond that. No clear answer. No smoking gun. Just bad fuel that made its way into a good machine.

Lessons Learned:

In aviation, trust is earned, but it should never be blind. I trusted that the fuel was clean because it always had been. I trusted the truck, the tank, the system. But somewhere in that chain, something failed, turning my engine into a glider and nearly turning me into a statistic.

Fuel checks can't *just* be visual. Foam, discoloration, or strange texture mean something's wrong. And if you don't find it on the first look, look again. And again.

The enemy isn't always mechanical. Sometimes, it's what flows through the veins of your aircraft. And if you're not vigilant, it can take everything down with it - quietly, invisibly, and instantly.

Fuel contamination doesn't just stop engines. It stops hearts, breaks machines, and rewrites confidence. So now, before I trust anything that goes in the tank, I make sure it earns it. Every time.

<u>NOTES:</u>

THE LEVER THAT LET GO
CESSNA 207

It was June, mid-morning, and I was already deep into the day's work - three flights done, the airplane running beautifully, and now loaded up again with another round of sterile fruit flies to release. I lined up for departure out of Harlingen and rolled the throttle forward in the Cessna 207. We took off clean, climbed to about 500 feet AGL, and I leveled off.

As I settled the airplane into cruise, I adjusted the throttle: 2,500 RPM and 25 inches of manifold pressure - just like always.

And then... nothing.

The power dropped instantly to idle. Not a cough, not a sputter - just silence and a sudden sense of disbelief. The engine hadn't quit - it had simply stopped responding. I pushed the throttle forward again, checked mixture, checked everything. No change.

There was no way I was making it back to the airport. I scanned the terrain and spotted a field. That was my only shot.

I put the nose down, kept my glide, and committed.

The landing was rough. The left wing struck hard and crumpled as we skidded through the uneven ground.

When we came to a stop, I sat there in silence, hands still on the controls, processing the fact that I was okay. Shaken - but alive.

After the wreck, we tore into the throttle system. What we found was surprising, and frustrating.

The throttle control lever had fractured - cleanly disconnected from the cable. The lever bore and collar were worn, splines uneven and chewed. The connection was so degraded it couldn't hold under normal pressure. Even worse, the hardware build-up on the throttle assembly didn't match the service manual. A washer was missing - just a simple washer.

That missing piece let the connection wear out slowly over time, until it failed completely.

This wasn't a sudden event. It was a slow-motion failure, developing over months, or years, until the moment it mattered most.

Lessons Learned:

Aircraft don't just speak through gauges and warning lights - they speak through wear. And when we fail to listen, they will shout.

This wasn't a blown engine or a fuel problem. This was a bolt, a washer, and a worn piece of metal that didn't get the attention it needed. The engine had been overhauled only 86 hours earlier. But someone, somewhere, didn't follow the manual exactly. Just one missing washer - and I ended up in a field with a broken airplane.

Maintenance isn't about fixing what's broken. It's about protecting what isn't. That means following the exact build specs. That means inspections aren't just paperwork - they're a shield between safe flight and sudden failure.

Now, I check the build-up. I check the small parts. Because I've learned that big accidents are often built from the tiniest omissions.

NOTES:

THE NUT THAT LET GO

BELL 206B JETRANGER

It was early August, and the job was simple: a low-level aerial application over a cornfield near Arcadia. The air was clear, visibility perfect, and I was flying just six feet off the deck. The Bell 206B was doing its job - light on the controls, responsive, humming smoothly. I was in the groove.

Until I wasn't.

Mid-run, I felt the engine cough. Then silence.

No warning, no gradual fade. Power dropped instantly, and there was no time to troubleshoot. I was already low. I aimed for a patch of crop that looked less dense and tried to flare out. But the descent was steep, and the rotor speed was bleeding fast. We hit hard, blades still winding down, skids punching into the dirt.

A flash of heat erupted behind me - a fire near the engine. I scrambled out, heart pounding, ears ringing. I was bruised and shaken, but alive. The fire didn't spread far, and someone nearby put it out before it consumed more than the cowling.

Later, investigators combed through the wreckage. No signs of fuel contamination or oil starvation.

All systems had been functioning. Except one. A single pneumatic line - known as the Pc line - had disconnected.

That tiny metal B-nut that holds the Pc line to the fuel control unit had backed off. It wasn't torqued correctly. No fresh torque paint, no safety marks, no lockwire. It had likely been missed during maintenance, just a few days and flight hours before the crash. A small oversight - on a hard-to-reach fitting.

That B-nut was vital. It allowed the engine to "talk" to the fuel control unit, adjusting fuel flow to meet power demands. Without it connected, the fuel control unit defaulted to idle. The engine didn't fail - it simply stopped trying to produce thrust. And the helicopter became a rock.

There had been an over-temp event a month earlier, and during maintenance they'd removed the turbine, swapped out the fuel control unit, and done all the inspections. But in that cluster of tubes and lines, one critical connection hadn't been secured.

Lessons Learned:

In aviation, the smallest parts can carry the biggest weight.

That B-nut - barely bigger than a coin - was the only thing between flight and free fall. It wasn't sabotage, it wasn't age, and it wasn't design. It was missed torque. A detail. A line item. Something easily skipped in the rush to button up a job.

This accident taught me that it doesn't matter how many hours you've flown or how perfect the day is - if something as basic as a fitting isn't double-checked, it can ground you faster than any storm.

Now, I check the maintenance logs like they're a map back home. I ask the questions. I trust - but verify. Because the next time something lets go, I may not be low over corn with a soft patch to land on.

A $2 nut nearly cost me everything. And I won't forget it.

<u>NOTES:</u>

THE POWER THAT FADED
THRUSH S2R-H80

It was early May, and I'd just loaded 325 gallons of mix for another routine aerial application flight. Everything about the setup was familiar - same strip, same aircraft, same early morning rhythm. The air was smooth, and I lined up on the runway expecting nothing more than another standard takeoff and turn into the field.

But the aircraft didn't lift off like it usually did.

I noticed it right away. We used more runway than normal, and even once airborne, something felt off. The engine wasn't pulling like it should. It was producing power, but not enough - not what I knew it was capable of. I started a gentle right turn, trying to ease into the spray run, but I could feel the airspeed dropping. It wasn't just a sluggish climb - it was a slow slide downward.

I scanned the gauges. One number jumped out: fuel pressure - fluctuating. I knew right then I had a fuel issue. I reached for the emergency pump switch and flicked it on, hoping it would kick in and stabilize things. It didn't.

There was no time to troubleshoot.

The trees were getting closer.

I jettisoned the load to try to buy myself some time.

Shed the weight, lighten the aircraft. But it was too late. The descent continued, and seconds later, we hit.

The crash was brutal. Trees tore through the wings, and the ground slammed the aircraft into a halt. But somehow, I walked away without a scratch.

Later, they tore the plane apart and found the answer: the electric main fuel pump had failed. Inside the pump, the carbon brushes were completely gone - worn to nothing. No material left in either brush holder. The electric motor had no chance of functioning. That explained the erratic fuel pressure and the loss of power. The emergency pump worked fine on the bench, but it hadn't been enough - or hadn't kicked in fast enough - to recover.

The aircraft manufacturer didn't have a replacement interval for the pump. No life limit. No overhaul schedule. As long as it operated and didn't leak, it passed annual inspection. There was no entry in the logs showing when that pump was last installed or serviced. All I knew was that the failure happened at the worst possible time - just as I needed power most.

Lessons Learned:

In aviation, "it's working" isn't the same as "it's reliable."

This wasn't a sudden catastrophic failure - it was wear, plain and simple. A set of brushes inside an electric motor that had quietly ground themselves into dust until there was nothing left. No alarms, no checklist items, just a slow decline into failure.

That accident taught me that preventive maintenance doesn't stop at what's visible or obvious. If a component can end your flight, it deserves attention - even if it's buried inside a pump that passed last month's inspection.

Since then, I've made it a point to ask more questions. To open panels that aren't "required." To think about time-in-service, not just whether something leaks or turns on.

Because in ag flying, when you're this low and this loaded, you only get one chance when something quits.

And from now on, I won't wait for performance to tell me there's a problem. I'll dig deeper. Because if you wait until power fades, it might be too late to pull up.

NOTES:

THE POWER THAT FROZE
PIPER PA-25

It was a crisp morning in June. I was in the Piper PA-25, engine running, waiting patiently at the end of the runway for oil temperature to rise and pressure to drop. It was my first flight of the day, and I wasn't rushing. I'd started up, taxied out, completed the run-up, checked magnetos, and even toggled the carb heat - just like I'd done hundreds of times before.

But while I sat there at idle power, waiting, the air around me was quietly working against me.

About ten minutes later, everything looked good enough to go. I lined up, rolled forward, and took off. It wasn't a long climb - just the normal circuit to get myself aligned before heading to the spray job. But as I turned onto the crosswind leg, I felt the engine hesitate... then fail. Just like that.

Silence.

No power. No time. I dropped the nose, scanned ahead, and aimed for a gravel road. It wasn't ideal, but it was the only option.

I came in fast and low, touched down straight, but I couldn't stop before the ditch.

The airplane slammed through it, wings digging into the terrain, bending like paper. The damage was immediate and brutal. But I was alive. Unhurt. Just stunned.

After the crash, we looked over everything. No mechanical failure. No bad fuel. Nothing wrong with the engine. That's when the weather data told the story: temperature around 75°F, dew point 62°F. A perfect storm for carburetor icing - especially at idle.

Turns out, I'd sat too long on the ground, waiting at idle power in just the wrong conditions. That cold morning metal mixed with warm, moist air, and ice slowly crept into the carburetor throat. I'd checked the heat during run-up, yes - but once I shut it off and sat there waiting... that's when the ice built up.

By the time I went full power for takeoff, the damage was already done. The airflow was restricted, the mixture thrown off, and the engine couldn't keep running.

Lessons Learned:

Carburetor ice isn't just a cold-weather problem.

It can sneak up on you in summer, on sunny mornings, when you're not thinking about it. And it doesn't need freezing temperatures - it just needs the right combination of humidity and throttle setting.

I did everything I thought I should have - but I didn't anticipate how quickly carb ice can form when sitting at idle. Waiting too long with the carb heat off created the perfect environment for induction icing. And once it forms, there's no climbing away from it.

From now on, I treat every idle wait like a potential trap. I stay on carb heat longer. I stay alert, even on warm days. Because now I know - what you can't see forming can still bring you down.

In aviation, silence isn't always peace. Sometimes, it's warning you that the engine's about to quit.

<u>NOTES:</u>

THE POWER THAT SLIPPED AWAY
AIR TRACTOR AT-502B

It was late in the morning, sometime in June, and the day was already heating up. I'd flown out of a narrow 2,100-foot airstrip plenty of times before - nothing fancy, just a solid stretch of pavement good enough for the work we did. My Air Tractor was loaded heavy: a full hopper of fertilizer, around 3/4 tanks of fuel, and a bit of density altitude creeping in with the heat. But nothing felt off. It was another job, another field, another takeoff.

Until it wasn't.

I was just a few hundred feet off the ground when I felt the thrust fade. Not a sputter or flameout - just a steady drop-off. The engine was still running, but I could feel the power bleeding away. The propeller shifted, moving toward a feathered position, and I lost thrust fast.

No time to troubleshoot. I started dumping fertilizer, trying to lighten the load. Altitude was vanishing quickly, so I picked the nearest field and committed.

I came in hot and low, the aircraft sluggish and heavy.

The forced landing tore through the field, smashing the fuselage, wings, and empennage. When we finally came to a halt, I just sat there. Everything had gone quiet. But I was unhurt. Shaken, but fine.

After the crash, we ran through everything.

The engine was still turning at the time of impact. And the propeller was damaged in a way that showed it was spinning, just at low power and pitch.

There was no catastrophic failure, no fuel leak, no system malfunction. The beta system worked fine post-crash. The propeller's feathering action had started, but no one could explain why.

I'd just had maintenance done the day before - something minor, a tweak to the beta plunger - but there was no logbook entry, no detailed record. And that's where the uncertainty began.

Fuel was another question mark. I'd reported having about 120 gallons on board. But when the FAA showed up, they found the tanks dry - no fuel smell, no spills, no leaks. Just empty tanks and more questions. Neither I nor the operator had removed any fuel, but it was gone. Maybe the jettisoned weight had something to do with it, or maybe the fuel burned off in the final run-up. No one could say for sure.

Still, the real issue wasn't fuel or engine failure. It was that mysterious drop in thrust. Everything looked fine mechanically. Nothing obvious. But the aircraft had simply refused to climb.

We reviewed the weight figures - over 8,000 pounds, maybe more. That runway, under those conditions, gave me maybe a few dozen feet of margin. Just enough to get off the ground, but not enough if something slipped. And something had.

Lessons Learned:

Sometimes, the failure isn't loud. It doesn't come with a bang or a warning light - it's a whisper. A subtle shift in power. A prop that feathers itself. A plane that flies, but just barely.

In this case, it wasn't pilot error, it wasn't an engine explosion - it was something undefined, some gremlin in the thrust system that robbed me of power at the worst possible moment.

But here's what I learned: heavy loads, short strips, and summer heat leave no room for doubt. If anything - anything - feels off, it's too much risk to take.

From now on, I double-check everything after maintenance, even if it's "just a quick adjustment." I verify logs, ask hard questions, and walk away from uncertainty. Because in ag flying, your margin for error is razor thin.

And sometimes, the thing that brings you down isn't what breaks. It's what slips, unseen, just enough to keep you from flying away.

NOTES:

THE SHIMMY THAT SPOKE
CESSNA A188B

It was August, and the heat was already baking the concrete at Chillicothe Municipal by midmorning. I was well into a solid rhythm - this was my fourth flight of the day, spraying fields just outside town in the old Cessna A188B. It had been a smooth operation so far, no signs of trouble. But as every pilot learns - sometimes it's the fourth flight that bites.

I came in for landing, lined up clean, and touched down with the precision that only repetition can bring. But just as the tail settled, I felt it - a shimmy. Subtle at first, then stronger. The tailwheel vibrated, and the aircraft pulled slightly to the right. I corrected with rudder and let it roll out. It wasn't dramatic, just...off. But once I slowed, everything felt normal again, and I taxied over to reload.

Did I note the shimmy in the back of my mind? Absolutely. Did I stop the aircraft, shut it down, and inspect the tailwheel? No. After three flawless landings and no prior signs of trouble, I chalked it up to surface bumps or wind. The temptation to keep moving is strong in ag work. The pressure to stay efficient. To do "one more run."

Once reloaded, I taxied out again, headed for takeoff. But the aircraft wasn't tracking straight. I had to use constant left rudder and brake just to keep her on the centerline. It was work - not impossible, but definitely not right. Still, I pressed on.

Lined up on Runway 14, I advanced the throttle. The engine roared to life, the tail came up, and the shimmy returned, worse than before. In seconds, I was fighting for control.

The aircraft veered hard right. I pulled power back, slammed the brakes, and tried to hold her straight.

Too late.

The airplane broke from the runway and skidded into the grass. As the right wheel dug into the turf, the tail whipped around, and we ground looped in a cloud of dust and humiliation.

When the dust settled and I stepped out, the damage was clear. The tailwheel had partially separated from the supporting structure. The whole empennage showed signs of overload stress, twisted by the forces of the loop. The tailwheel itself was bent hard to the right - proof it hadn't failed cleanly in the accident. It had likely been compromised earlier - probably on that last landing.

Later, during the inspection, they found no preexisting mechanical anomalies - no hidden cracks, no material flaws. Everything pointed to one thing: the tailwheel had been damaged during that previous landing, and I had missed the warning signs. Or worse - ignored them.

I'd felt it. The aircraft was talking to me, and I chose not to listen.

Lessons Learned:

Airplanes always speak before they scream.

Every shimmy, every vibration, every off-center roll is a whisper from the machine that something isn't right. And too often, pilots convince themselves it can wait - just one more load, just one more takeoff.

But anomalies don't fix themselves. In taildraggers especially, a compromised tailwheel isn't just a minor nuisance - it's a directional control system on the brink. Once it's gone, you're a passenger on the ground, not a pilot.

That day taught me that the most dangerous failures aren't sudden - they're the ones you notice, rationalize, and proceed with anyway. You don't always get a second chance to listen.

When something feels off, stop. Check. Ask the question you're hoping doesn't need an answer.

Because one shimmy can speak louder than a crash.

NOTES:

THE SILENCE AFTER THE TURN
GARLICK OH-58A

It was a typical summer afternoon in Georgia - hot, humid, the air alive with insects and turbine blades. I was flying a Garlick OH-58A+, a military helicopter repurposed for agricultural work. After an early morning departure from Greensboro, I'd flown up to Cisco for the day's mission. On arrival, I met with the landowner, walked the field, and even did a short orientation flight to survey the terrain.

The helicopter stayed running. Everything looked good. This was my routine - strap in, spray, reload, repeat.

The first load of 40 gallons went off without a hitch. The second, 50 gallons, felt just as smooth. I refueled, left the engine running, kept my harness tight. The turnaround was quick. Normal. Just another day.

When I lifted off with the third load - another 50 gallons - I felt the extra weight. It was noticeable, but still within limits. All my instruments were in the green. The engine sounded strong.

I eased into a teardrop turn, lining up for the next pass.

Then everything changed.

Mid-turn, the engine coughed. RPM dropped.

The red ENGINE OUT light blinked to life. My gut dropped with it. There wasn't time to think - just react.

Ahead of me was a narrow gap in the trees. No open fields. No smooth landings. Just dense forest and a slim chance. I lowered the collective and committed to an autorotation.

The descent was fast. I aimed for the gap, hoping the skids would hold. We hit the ground - hard. One skid caught first and flipped us forward. The aircraft rolled onto its side. The rotor shattered against the trees. The tail boom snapped.

And then - nothing. Just silence.

I was still strapped in. Disoriented, but breathing. The cabin had held. No fire. No smoke. Somehow, I was okay. I climbed out into the wreckage, shaken but unhurt.

When investigators arrived, the helicopter was barely recognizable. Main rotor blades fractured. Tail rotor destroyed. Torque tubes and drive shafts twisted and torn. It looked like the aftermath of something far worse than what I walked away from.

I told them exactly what I'd experienced. There had been no warning, no signs of trouble. Fuel was good. Performance had been solid all day. Then - nothing. Just a cough, a drop, and an engine out.

The engine was removed and shipped off for testing. To everyone's surprise, it ran. The only issue found was damaged compressor blades. Once replaced, the engine performed at all settings without fault.

There was no fuel contamination. No system obstruction. No mechanical failure they could confirm. The final report stated it plainly: a total loss of engine power for undetermined reasons.

Lessons Learned:

This one left more questions than answers. And that's the point.

Sometimes, engines fail and leave no clue. No broken rod. No burst hose. Just silence. And when it happens, it's not the red light that saves you. It's how fast you respond.

I survived because I didn't wait. I recognized the failure the second it began. I'd run through this emergency in my head hundreds of times. I'd looked at that terrain and memorized the only way out, even if it was ugly.

That's the game in ag flying - especially in helicopters. You're low. You're heavy. You're alone. And your margin for error is inches wide.

So here's what I learned:

- Don't rely on diagnostics to tell you what you already feel.
- Practice autorotations until they're reflex.
- Scout every escape route like you'll need it today.
- Expect the failure. Even when the numbers look fine.
- And if something feels off - even slightly - trust it.

I didn't bring the helicopter home. But I brought myself back in one piece. And when the engine goes silent mid-turn, that's the only result that matters.

Fly ready. Every time. Because the engine doesn't always warn you. And when it stops, you only get one chance to get it right.

<u>NOTES:</u>

THE TAKEOFF THAT DIDN'T TAKE
GRUMMAN G-164B AG CAT

It was a warm June afternoon in Nebraska, and I was ready for the second flight of the day in the Grumman Ag Cat. The first flight had gone smoothly - engine strong, conditions clear, and the airplane feeling solid. No signs of trouble. I refueled, reloaded, and taxied out, going through the same checks I always did. I lined up on the strip, confident and ready to go.

Throttle in. Airspeed up. The tail came up, and the airplane began to lift - but something wasn't right.

We broke ground, but just barely. I was expecting a climb, a clean arc into the sky - but the aircraft just hung there, refusing to gain altitude. I jammed the throttle forward, trying to coax more power, but the engine didn't respond. Not a surge, not a cough - just nothing.

The end of the runway was closing fast. A fence loomed ahead. There was no lift, no climb, and no time.

We punched through the fence and slid into the soybean field beyond, the airplane smacking hard onto the dirt. The right wings - both upper and lower - took the worst of it. Crushed. Twisted. But the cabin held. I was rattled, but unhurt.

After the crash, we dug into the engine, trying to find answers.

There was no explosion, no catastrophic failure. The propeller blades were chewed and deformed, as you'd expect from impact with the ground. The engine showed signs of power on contact - rotational scoring on the turbines and compressor impellers, earthen material sucked into the core, even a torsion shaft fractured from sudden stoppage.

It had been producing some thrust. But not enough.

We bench-tested the fuel control unit. At first, the cam moved as it should when pressure increased - but then it stopped. Locked. Disassembly revealed a seized cam gear, worn bushings, and scarring along the shaft. One gear tooth was cracked end-to-end.

It was damage, yes. But was it the reason?

Unlikely. The experts said that if the camshaft had been binding, it would've caused issues earlier - on the previous flight, or even before that. Plus, there was Jet A fuel and some water in the fuel sample, but not enough to explain a power loss.

The engine hadn't failed dramatically. It had simply... underperformed. And despite all the investigation, we couldn't pin down why.

Lessons Learned:

Sometimes the scariest failures are the ones that don't leave clues.

We expect engines to quit loud - to bang, to flame out, to leave a trail of wreckage that tells the story. But this wasn't that kind of failure. This was a slow fade. A silent refusal to give full power. The kind of malfunction that slips past your checks and only shows itself when you need everything working.

That's what makes it dangerous.

It reminded me that flying agricultural missions at max weight, in high density altitude conditions, is like walking a tightrope. There's no room for performance margins to shrink. If you're heavy and the engine isn't giving you everything it has, there may not be a fence at the end of the runway - just a wall.

In the end, I walked away. But the airplane didn't.

And now I fly every takeoff like the engine might not deliver. Because someday, it might not - and when it doesn't, the only thing left is how you respond.

NOTES:

TORN FROM THE SKY
CESSNA A188B

It was a hot afternoon, and I was just half a mile from the strip, flying a load of dry fertilizer toward a nearby cornfield. I was in a machine I knew well - an older ag aircraft that had served me faithfully for years. I had over 4,000 hours flying, more than a third of them in this model. Conditions were ideal. The sky was clear, the wind calm, and everything had passed inspection just weeks earlier. From the outside, it was a normal run.

Inside, that changed fast.

As I dropped low over the field to begin the pass, I felt something shift. It wasn't dramatic - no bang, no violent shudder - just a slow, unsettling loss of power. I pushed the throttle fully forward. Nothing changed. I flipped on the auxiliary fuel pump. Still nothing. The engine was running, but it was too weak to stay airborne at low level.

I scanned the terrain. No roads, no fields I could trust. Trees and wires were all around. I had only seconds.

I aimed for the largest gap I could find and committed to a forced landing.

As I descended, the aircraft clipped trees, then wires. I felt the wings rip against the branches. The fuselage strained as I hit hard.

I stayed upright, but then came the fire. Flames burst from the fuselage, licking up around the cockpit. I was stunned but alive, and I didn't wait - I scrambled out fast, ran clear, and turned to see the wreck that had been my aircraft engulfed in flames.

The airplane was gone. But I was still standing.

Investigators later combed through the remains. The cockpit was destroyed. The fuselage burned through. But enough of the engine remained to tell part of the story. In the fuel manifold assembly, they found it: a small tear - just about an inch long - in the diaphragm. That tiny rupture had quietly degraded fuel flow until the engine couldn't maintain power.

There had been no warning signs. No leaks. No fuel system red flags in the logs. The aircraft had passed inspection. The engine hadn't been overhauled in years - but it had seemed fine. This fault was invisible until it mattered most.

The conclusion was clear: partial power loss caused by internal failure of a key fuel system component. There had been no pilot error - just a mechanical failure hiding in plain sight.

Still, the outcome could've been worse. I'd flown low, fast, and into danger like we all do in this business. But when it went wrong, I reacted fast, aimed for a survivable spot, and got out when it counted.

Lessons Learned:

Not all failures scream before they strike.

This one didn't. It crept up quietly, the result of age and fatigue in a part buried deep in the system.

Here's what I learned - and what others should, too:

- Fuel systems can hide flaws even after passing inspection.
- If your engine hasn't been overhauled in more than a decade, don't assume it's healthy just because it runs.
- "No squawk" doesn't mean "no problem." Parts wear silently - especially ones that deal with pressure and flow.

- Pre-flight prep has to go beyond the obvious. Think about component life, not just airframe hours.

I lived through this because I reacted fast - but I was also lucky. And luck should never be part of the safety plan.

Ag flying pushes every limit - altitude, airspeed, time, terrain. There's no margin for guessing. If your engine gives you even the hint of hesitation, you won't get time to wonder why. You'll get one chance to survive it.

I got mine.

Next time, I want to make sure it's never needed.

NOTES:

UNRESPONSIVE POWER
AIR TRACTOR AT-502B

The sun had just started to dip behind the Kansas horizon when I lifted off for another routine spray run. Golden fields stretched endlessly beneath me, the skies were calm, and everything about the evening felt predictable. No radio chatter, no traffic - just the hum of the Air Tractor AT-502 and the familiar rhythm of agricultural flying. It was my second swath run. Everything felt routine. Until suddenly, it wasn't.

As I began my turnaround maneuver, something shifted. The tail dropped without warning. At the same time, every engine gauge on the panel - except one - spiked past redline. Oil pressure, torque, RPM - all surging. Only the Interstage Turbine Temperature held steady.

I reached instinctively for the power lever, trying to throttle back or find some control. Nothing. No change. The lever moved, but the engine didn't care. It was running wild, completely unresponsive to commands. In that moment, there was no checklist to follow - no neat flow of boxes to tick.

There wasn't time to panic. Experience kicked in. The aircraft had to get down, and fast.

I scanned the terrain. At low altitude, with trees and power lines dotting the landscape, options were limited. But a gravel road cut through the farmland - a narrow, winding path, but it was something.

I lined up the aircraft as best I could. The trees at the edge of the field were close - closer than I liked - but there was no turning back. As I descended, the wingtips started brushing branches. I could feel each strike vibrate through the controls. A reminder of just how tight this landing was going to be.

Then, wheels hit gravel. Hard. The aircraft slammed down and skidded, fighting for traction. I tried to hold it straight, but it veered, then spun. The airplane ground-looped into a ditch, and everything came to a jarring halt.

Inside the cockpit, I was still strapped in, engine roaring full throttle. It wouldn't idle down. I reached for the Start Control Lever and yanked it into "Fuel Cut-Off." The noise stopped. Silence returned. The aircraft was broken - but I was alive.

I climbed out slowly, heart racing, legs shaky. The wings were mangled, folded in on themselves from the impact. The landing gear was a mess. But there was no fire. No fuel leak. No explosion. And somehow, not a scratch on me.

The field crew arrived quickly. We secured the area and documented everything. Later, the investigators would comb through the systems, trying to figure out why the engine ignored the throttle. As of then, it was still a mystery.

What I did know was this: clear skies and calm winds don't guarantee a safe flight. And even the best-maintained machines can suddenly go off script.

I'd never experienced a failure like that before - where the engine wasn't out, but also wasn't in my control. It was like flying a runaway train. The controls were there, but none of them mattered. The only system I had left was my brain.

Lessons Learned:

We spend hours in training going over procedures for engine failure, power loss, mechanical trouble - but real failures don't always play by the book. This wasn't a textbook emergency. There was no clean checklist for "engine runs wild and won't listen."

In that moment, I fell back on something deeper: judgment, decision-making, and calm. I remembered what an old instructor once told me - fly the airplane as far into the crash as possible.

I didn't panic. I flew. I kept flying. I picked a road, avoided obstacles, and got it down.

Machines can fail without warning. But keeping a clear head can turn a worst-case scenario into a survivable one.

When everything else goes silent - or, in this case, deafeningly loud - your mind might be the only reliable tool you have left.

NOTES:

WHEN SPARSE ISN'T SO SIMPLE
UNSPECIFIED AGRICULTURAL AIRCRAFT

It was a routine morning in eastern North Carolina. I had just launched from a strip near Tarboro to apply cotton treatment to fields located just east of the local airport. The conditions were perfect with clear skies, smooth air, and familiar terrain. I had flown this area many times before. With 6,000 hours in my logbook - half of that in agricultural ops - I was confident in my technique and judgment.

The flight complied with all FAA Part 137 and North Carolina pesticide application regulations. The fields I was treating sat within rural terrain. There were a few homes scattered about, and one modest trailer park with fewer than ten trailers. This was standard for the kind of areas we often service - areas I would define as sparsely populated, and well outside any town center or congested area.

But later that day, a phone call changed the tone of the morning. A resident had filed a complaint.

Suddenly, what I had considered a routine mission became a regulatory grey area. Was this area sparsely populated, or did it meet the FAA's unclear threshold for congested? That distinction matters a great deal in ag flying.

Under Part 137, flying closer than 500 feet AGL over sparsely populated areas is permitted. But if the FAA determines the area to be congested, even a routine spray job could become a violation unless advance planning and FSDO notification are completed.

In this case, I was within the limits - by both distance and regulation, as I understood them. But the lack of FAA clarity around what qualifies as "congested" leaves aerial applicators guessing. Was the small cluster of trailers enough to tip the scale? I didn't think so. But when complaints come in, it's often the interpretation that matters, not the intent.

Lessons Learned:

This experience reinforced that precision in ag flying goes beyond airspeed and altitude - it includes interpreting vague regulations.

The FAA's definition of a "congested area" remains unclear, leaving aerial applicators vulnerable to differing interpretations, especially when a complaint is lodged. What seemed like a sparsely populated area to me - a few scattered homes and trailers - was enough to trigger concern and scrutiny.

Now, if I have even a hint of doubt about a location, I take the extra step of contacting the FSDO beforehand. It has added time, but worth the peace of mind. Because in low-level ag operations, perception can override intent. And prevention beats investigation every time.

NOTES:

PART 4 - NEAR MISS & SEPARATION

"When pilots experience a near miss with another aircraft, or have been flying at the wrong altitude, they file a report. ... You can't live long enough to make them all yourself."
  –Matthew Syed

A THIN COAT OF PAINT
UNSPECIFIED AGRICULTURAL BIPLANE

It was a routine departure from Mount Vernon Airport (MVN). The skies were clear, the load was heavy, and I was lined up on Runway 33 in my big radial-engine ag biplane, ready for takeoff.

As I began the roll, focused on keeping the aircraft straight, something in my peripheral vision yanked me from routine into reflex. A blue pickup truck - moving fast - was crossing the active runway ahead. I yanked the stick, braced, and somehow - miraculously - we didn't collide.

Witnesses later said we missed by what looked like nothing more than a thin coat of paint. I still don't know how we didn't make contact... though later inspection showed the tip of my wing had clipped the camper shell of the truck.

Once safely on the ground, I confronted the driver. He was a construction worker, in a rush to get to the other end of the runway to remove an "X" for an incoming airline flight. He admitted he hadn't looked before crossing. I was furious. I told the airport needed a flagman - immediately.

An hour later, I was back at the end of the runway, loaded again.

This time, a blue FAA van cruised across the same runway without stopping. No caution. No radio call. Nothing. I was fully visible - revving at the hold point. And still, another near miss.

I called the FAA to report the near-collision. They told me the airport manager and the city were responsible. I spoke with the manager, who promised a flagman would be present the next morning.

That never happened.

Since then, trucks - including that same blue pickup - continue to cross the runway daily without stopping or checking for aircraft.

In 34 years of flying, I've never seen anything quite like it. And unless something changes at MVN, it's not a question of if something happens - but when.

Lessons Learned:

This close call drilled home a simple truth: vigilance doesn't end at the propeller's edge. Runways aren't just shared with airplanes - they're magnets for vehicles, especially at airfields lacking proper oversight.

I learned never to assume anyone else is scanning for traffic, even on an active runway. From now on, I treat every takeoff like a blind intersection - eyes wide, reflexes ready. It shouldn't be this way, but when flagmen are absent and communication is nonexistent, your best defense is expectation. Expect carelessness. Expect ignorance. And plan for both.

Safety depends not just on procedures, but on a mindset: assume nothing, anticipate everything, and stay ready to dodge what no one else sees coming.

NOTES:

BLIND CORNER MIDAIR
UNSPECIFIED AGRICULTURAL AIRCRAFT

I was working a cluster of fields just north of Munday, Texas. My routine was pretty typical - flying east-west spray runs from the north end of the fields, gradually working my way south.

After completing the final pass, I lined up for a clean-up swath to tidy up the southern edge of the field. That heading pointed me straight toward town, so as soon as I finished the pass, I pulled into a steep climb and made a sharp westbound turn to avoid flying over any populated areas.

I had just rolled out on a heading of 270°, cruising at about 475 feet AGL, when my stomach dropped. Out of the corner of my eye - dead ahead - an ag aircraft was coming straight at me from the west.

Same altitude. Same airspace. No warning. No time.

I yanked the stick left and dove southwest, banking hard to avoid a midair. The other pilot never flinched. No change in heading. No visible reaction. Just kept cruising straight through the airspace I had just cleared with a steep climb and turn.

We missed by maybe 100 feet laterally - zero vertical separation.

Lessons Learned:

The moment I broke above the treeline, I thought I was clear. I'd just finished a routine spray run and made a sharp climbing turn away from town, assuming the sky ahead was mine. It wasn't. Another ag plane was barreling straight toward me, same altitude, no warning, no reaction.

We missed by what felt like inches. In those low-level ops, we tend to fixate - on GPS bars, terrain, swath width - but the real threat might be just over the next ridge or turning in from the opposite side.

From now on, my pull-ups will include a deliberate scan - left, right, above, and behind. Because even at 475 feet, the sky can feel like a blind corner.

And one unchecked assumption can bring two aircraft terrifyingly close to becoming one.

NOTES:

CHASED THROUGH THE FIELDS
UNSPECIFIED AGRICULTURAL AIRCRAFT

It was just another spring day in southern Illinois - perfect weather, 2000-foot ceilings, and about 7 miles visibility. I was sowing soybeans into wheat fields just north of Carbondale, working off plat books with flaggers marking the rows below.

As the day went on, my flagger became harder to spot and I found myself drifting farther south. Eventually, I located him again, lined up, and began spreading. That's when I saw it - out of the corner of my eye, a Cessna 152 about 1000 feet to my right.

A second later, I looked up and spotted the Carbondale airport (MDH) about a mile to the south-southwest. That's when it hit me - I had just slipped into Class D airspace without contacting the tower.

I wrapped up the run, performed my end turn, and began heading back north toward my home base at PJY at about 200 feet AGL.

Then it got worse.

Another Cessna appeared - this one just 200 feet off my right wing, flying in the same direction. I banked slightly left to open the distance, and the other plane eventually veered off.

Back at PJY, we phoned the tower to own up.

They told us they'd already reported the incursion to the Springfield FSDO - but if we called them before and after operating in their airspace, they'd allow us to complete the job. We did just that, with full cooperation from the tower on the remaining days.

But then the farmers showed up. They told me someone had been "chasing" me with a Cessna through the fields trying to get my tail number. That pilot had apparently received permission from the tower to get airborne and pursue me for identification.

That left me shaken. Yes, I had made a mistake - I entered controlled airspace without a clearance. But the idea that another pilot was tailing me, unannounced and low over active spray runs, was not just reckless - it was dangerous.

I've flown 21 seasons without a violation. I admitted my error, and I'll work harder to make sure it never happens again. But chasing an unsuspecting aircraft through a field? That wasn't the way to handle it. Two wrongs don't make a right.

Lessons Learned:

This incident taught me that controlled airspace boundaries can sneak up fast - especially when you're focused on the job and relying on ground flaggers or plat maps. But the bigger shock came afterward: being pursued at low altitude by another aircraft trying to identify me.

Yes, I entered Class D airspace without clearance, and I took responsibility for it. But chasing someone down at 200 feet AGL adds danger, not resolution.

In ag flying, safety must be a two-way street. From now on, I'll double-check sectional charts before every mission - and I'll advocate that conflicts be resolved on the radio or the ground, never through airborne confrontation.

NOTES:

CLEARED TO COLLIDE?
THRUSH S2R-T

During an aerial application operation, I was taking off from Runway 08 at BWC. My aircraft, which lacks a mounted VHF radio, made no radio calls and did not hear any radio transmissions.

The light winds favored the use of Runway 08, and there was a general aviation fly-in at the east end of the airport. I taxied from the mid-field load pad on the south side of the runway, ensuring that my wingtip strobe lights and landing light were on to enhance visibility.

Upon reaching the entrance to Runway 08, I checked the traffic pattern for aircraft, saw none, and looked down the runway, where I observed no other aircraft on the runway or at the hold-short lines.

I taxied onto the runway and began my takeoff roll. As I was gaining speed, I noticed another aircraft starting its takeoff roll on Runway 26 at the opposite end of the runway. I elected to continue my takeoff, as I was nearing lift-off speed and knew I would be airborne before the other aircraft could reach me.

After lifting off, I maneuvered to the right of the runway (left of the parallel taxiway) and continued my climb. As I looked down and to the left, I saw the other aircraft still on its takeoff roll.

At an altitude of approximately 200 feet, I began a left crosswind turn. By 400 feet, I turned downwind, continuing my climb to 500 feet while heading west to spray a field. I noticed the other aircraft, still in a right crosswind turn, eventually exiting the pattern to the north. At this point, I saw another aircraft on a wide left downwind for Runway 08. To avoid conflict, I exited the pattern straight out from the downwind.

Several factors contributed to this event:

1. The other aircraft did not use the runway favored by the prevailing wind, opting for a slight tailwind instead.
2. The other aircraft did not visually check the runway before entering. Had they done so, they would have seen my aircraft's 650-watt landing light.
3. My aircraft did not have a VHF radio, but I believe radio communication should be mandatory at public airports to ensure clear coordination.

Lessons Learned:

This incident drove home how vital clear communication and visual awareness are - especially at airports with mixed traffic and no mandatory radio use. The absence of a VHF radio in my aircraft and the other pilot's decision to depart with a tailwind created a high-risk scenario. Visual checks must be thorough, and aircraft should always verify a runway is clear before entering, regardless of traffic or fly-in activity. Radios should be mandatory at public-use airports; without them, assumptions can become hazards. From now on, I'll advocate for clearer coordination protocols, and I'll never assume others see me - even with lights on. Vigilance is the only guarantee.

NOTES:

HEAD-ON AT TAKEOFF
UNSPECIFIED AGRICULTURAL AIRCRAFT

It was a calm, clear morning at the airfield, just like any other spray mission. I had just loaded up with a heavy batch of insecticide and was taxiing out for takeoff on Runway 11. The wind was light out of the north - about 5 knots - and all previous traffic had been using 11, so I followed suit.

My radio only had receive capability on 122.8, and I heard nothing. No calls. No traffic. Nothing. The airport was under heavy construction on the parallel taxiway, and between the dust, dirt piles, and equipment, visibility across the field was terrible - especially to the east end, about 5,000 feet away.

As I began my takeoff roll, tail lifting and speed building, I suddenly saw it: another aircraft, a small high-wing rental, barreling toward me down Runway 29 - head-on.

We were both committed. I held right, keeping my wheel almost scraping the runway's edge. He did the same. We passed within 20 feet of each other, each probably holding our breath.

No one got hurt. No metal got bent. But it could have gone very differently.

Later, I learned there were three aircraft on the east side of the field I never saw - hidden behind construction equipment. Apparently, someone had decided to switch runway direction without communicating it clearly, and the FBO didn't announce any change or answer incoming calls.

The truth is, this was a long-simmering safety issue. The FBO had been in decline for years: aging staff, financial struggles, and poor communication. The city's part-time airport manager couldn't do much. And to top it off, the shared UNICOM frequency (122.8) was overloaded by traffic from nearby Greeley, making it hard to communicate even when someone did try.

Lessons Learned:

At uncontrolled airports - especially those under construction - assumptions can kill. I launched on Runway 11, only to meet another aircraft head-on departing 29. Neither of us had full visibility, and poor field communication made things worse. My aircraft only received on 122.8, and with the local UNICOM cluttered by nearby airports, vital updates never got through.

The root problem wasn't just one bad decision - it was a system breakdown: unmonitored frequency, uncoordinated operations, and an FBO lacking leadership.

In high-stakes environments, assume others can't see you or hear you. If something feels off, stop. Confirm. And live to fly another day.

<u>NOTES:</u>

LIGHTS OUT AT DUSK
GRUMMAN G-164B AG CAT

It was a late July evening, and I was preparing for my second run of the night, departing Walla Walla Airport (ALW) in my G164B Ag-Cat. The tower had already closed for the day, so the airfield was running uncontrolled. With dusk settling in and a time-sensitive chemical mix in the hopper, the pressure was on to get the job done.

Earlier, I'd taken off with tower clearance using Runway 20, but this time, I opted for Runway 2 at intersection C to save time and reduce noise exposure over nearby residential areas. I made a clear radio call: "AgCat taking Runway 2, intersection C, right turnout." Everything seemed quiet - no traffic in sight or on frequency.

As I began my takeoff roll and straightened out onto the runway, a sudden movement caught my eye: a Cessna 152, lights off and coming straight at me, taking off from Runway 20 - opposite direction. We were head-on.

Both of us reacted quickly. I veered to the west side of the runway while he shifted east. We passed somewhere between intersections B and C, close enough for comfort but thankfully clear of contact. Then, we each continued our separate departures.

It's still unclear whether the Cessna had made any calls; I hadn't heard one. If he'd been in the pattern and made a long final call 2–3 miles out, I could've missed it while troubleshooting the flagger issue back at the hangar. Regardless, he wasn't showing any lights - and at that hour, it made him almost invisible against the dusk sky.

I'll admit: I should have double-checked the runway more thoroughly before rolling. But in uncontrolled airspace, the burden to "see and avoid" is a shared one. When someone's flying without lights and not broadcasting position reports, it becomes a dangerous guessing game.

Lessons Learned:

Dusk flying demands heightened vigilance - especially at non-towered airports. This incident drove home how quickly things can turn dangerous when one aircraft launches unlit and silent.

Even with a proper callout, I should have paused longer and visually confirmed both ends of the runway. In fading light, assumptions are risky and habits can cut corners. Just because the radio is quiet doesn't mean the airfield is empty.

From now on, I treat every uncontrolled field at dusk as high-alert. Lights on. Radios active. Runway checks doubled. Because when visibility slips and timing tightens, it's not just about doing the job - it's about making sure we all fly home.

NOTES:

MISSED CALL, MISSED SEPARATION
UNSPECIFIED AGRICULTURAL AIRCRAFT

It was a clear morning as I prepared to depart from with a full load in my Part 137 aircraft. Two other aircraft were active in the pattern: a Cessna 172 executing touch-and-gos on Runway XY and a Bonanza entering downwind for the same runway. My departure would be from Runway XX, intersecting their active pattern.

I made my intentions known over the radio: taxiing into position on XX and planning to wait for the Cessna to pass before I began my takeoff roll. I heard the Cessna acknowledge my call, and what I thought followed was a transmission that they were on short final and would hold short of XX. Believing I had a clear path, I announced my departure and began my takeoff roll.

As I accelerated down the runway and approached the intersection with XY, my heart jumped - out of nowhere, the Cessna appeared overhead, having just lifted off from its touch-and-go. We crossed paths with only about 150 feet of vertical separation.

After the flight, I called the Cessna pilot to discuss what happened. He explained that he had never said he would hold short; rather, he intended to do a quick touch-and-go to be clear of XX in time for my departure.

Somewhere in our exchange, I had misunderstood his transmission - and critically, I never had visual contact with his aircraft when I initiated my takeoff.

Lessons Learned:

In uncontrolled airspace, the radio is a vital tool - but it's no substitute for eyes on target. That morning, I made a critical assumption based on a transmission I thought I heard, not one I confirmed. The result? A dangerously close crossing with a Cessna during my takeoff roll - just 150 feet of vertical clearance between us.

The real error wasn't in the radio traffic - it was in relying on it as my only source of situational awareness. Without visual confirmation of the Cessna's location, I launched into a blind intersection, assuming the path was clear. It wasn't.

What this reinforced for me is something every pilot learns early on but must be reminded of often: never trust the radio alone. Aircraft can be misheard, misunderstood, or unseen - but your eyes, when used properly, don't lie.

No matter how busy the pattern or how clear the sky, never trade speed for certainty. If you don't have visual, don't go. Because in aviation, assumptions don't just make mistakes - they create emergencies.

NOTES:

MISHEARD, MISJUDGED, MISSED HIT

UNSPECIFIED AIR TRACTOR

The morning started like any other in Artesia, New Mexico, clear skies, good visibility, and a full day of aerial application work ahead.

The Air Tractor was fueled, loaded, and ready for another round of agricultural spraying. I had been running operations all morning off Runway 21, using the common UNICOM frequency of 122.8 for traffic advisories. After the FBO opened, I coordinated additional traffic updates with them, ensuring awareness of any incoming or departing aircraft.

At around mid-morning, I taxied out once again, preparing for departure. The aircraft was fully loaded for the next round of work. As I positioned on Runway 21, I heard a pilot in Aircraft Y report that he was on final for the crossing Runway 12.

I watched him land and clear the runway, confirming that the intersection was momentarily clear.

With the area seemingly safe, I announced my departure on 21 and advanced the throttle. As the tail came up, my eyes quickly scanned the surroundings. That's when I saw a light twin, Aircraft Z, just touching down on Runway 12, heading directly toward the intersection where our two paths would converge.

There was no time for hesitation. I aborted the takeoff immediately, pulling power and applying heavy braking. The right brake overheated under the strain and faded, causing the aircraft to veer left, off the runway and into the adjacent pasture. I fought to keep it straight, finally coming to a stop just after crossing onto Runway 12, about 500 feet from the intersection.

By then, the twin had slowed considerably and was just about to exit the runway normally. Over the radio, I heard the irritated voice of its pilot complaining about "the crop duster." As I taxied back to the ramp, I knew I needed to have a conversation with him.

When we spoke, his frustration was evident. He insisted that he had made multiple position calls on UNICOM and had never heard a response from me. Yet, I had been making my own calls, loud and clear - or so I had thought. To troubleshoot, we tested my aircraft's radio against a handheld unit and with the FBO's UNICOM receiver. Everything checked out fine.

The near-miss left no physical damage beyond my worn brake pads, but the risk of collision had been very real. I estimated our separation at about 750 feet - far too close for comfort. Without my evasive action, the outcome could have been catastrophic.

Lessons Learned:

- Radio Communications Are Not Foolproof – Even when both pilots believe they are transmitting and receiving correctly, technical issues or interference can still disrupt transmissions. This incident highlighted the importance of backup verification, such as listening for responses and double-checking frequencies.
- Visual Traffic Scanning is Critical – While I had made radio calls and monitored incoming aircraft, I should have conducted one final sweeping scan before committing to takeoff. Early detection could have allowed me to delay my departure and avoid the entire situation.

- Aborting a Takeoff Requires Precision – Heavy braking on an overloaded aircraft can lead to brake failure, directional control issues, or even a runway overrun. Knowing the aircraft's braking limitations in various load conditions is crucial.
- Uncontrolled Airport Operations Demand Extra Vigilance – With multiple aircraft using intersecting runways, extra caution is necessary. Pilots must be proactive in confirming the location
- of other traffic, especially when working in high-activity areas like agricultural operations.

Ultimately, this experience reinforced that in aviation, assumptions can be dangerous.

I walked away from the incident with a renewed commitment to thorough traffic scanning, more proactive communication verification, and a deep respect for the split-second decisions that can mean the difference between a lesson learned and a tragedy avoided.

NOTES:

NEAR MISS ON PIPELINE PATROL
UNSPECIFIED AIR TRACTOR & UNSPECIFIED PATROL AIRCRAFT

I was conducting a low-level pipeline patrol flight at 300 feet AGL under an FAA low-level waiver. As I proceeded along my route, a turbine-powered Air Tractor was performing aerial application nearby. Initially, our courses didn't intersect, but after completing a run, the Air Tractor pulled up and turned toward my flight path.

Seeing the developing conflict, I made a slight right turn and descended to maintain separation. However, the Air Tractor initiated another spray run, now on an apparent collision course with me. Closing in at a much higher ground speed, the agricultural aircraft was moving from my 4 o'clock position to my 2 o'clock position, slightly below.

Realizing the imminent danger, I executed an aggressive evasive maneuver, pulling up and banking hard right. We passed within 250 feet horizontally and just 40 feet vertically. Had I not reacted in time, our aircraft could have come dangerously close - perhaps within 50 feet of each other.

Upon reflection, I suspect that the ag aircraft never saw me.

Despite flying a white and blue aircraft that should have contrasted against the green fields, the other pilot appeared unaware of my presence. The unpredictable nature of agricultural flight paths, combined with the absence of effective strobe lights on my patrol aircraft, created a hazardous scenario.

Lessons Learned:

- Maintain Extra Vigilance in Uncontrolled Airspace: Agricultural aircraft and other specialized operations may not follow standard traffic patterns or CTAF procedures. I always assume additional traffic could be present and uncommunicative.
- Use All Available Tools for Traffic Awareness: MFDs, ADS-B, and onboard collision avoidance systems (if available) provide valuable situational awareness. However, I know they should never replace an active visual scan.
- Proactive Communication is Key: I announce position reports frequently, even when no other aircraft are heard, to increase awareness for others who may not be transmitting.
- Expect Unpredictable Movements from Agricultural Aircraft: Crop dusters operate at low altitudes, may execute rapid turns, and can be difficult to see. I assume they may not see me.
- Enhance Aircraft Visibility: High-intensity strobes, landing lights, and high-visibility paint schemes improve my chances of being seen, especially at low altitudes.
- React Decisively When a Conflict Arises: This case demonstrated how quick, decisive action prevented a collision. Having an escape plan in mind at all times can be the difference between a near-miss and a disaster.

- Advocate for NOTAMs When Necessary: If an airport
  has ongoing operations that significantly alter normal
  traffic patterns, I request NOTAM issuance through
  proper channels to help alert other aviators.

By internalizing these lessons, I can improve my awareness and
response strategies, ultimately ensuring safer skies for everyone.

NOTES:

NO RADIO, NO EXCUSES
PIPER PA-25 PAWNEE

During a return flight to the Airport after completing an agricultural mission, I encountered a runway incursion with a PA-28. Since my aircraft lacks a radio, I rely heavily on visual traffic spotting for awareness, which can sometimes be more challenging. On this occasion, I entered the traffic pattern for Runway 07 and completed a standard approach without seeing any traffic, except for the PA-28, which was taxiing for departure on the opposite Runway 25.

I continued my landing on Runway 07 and began braking after touchdown. Just as I was rolling out, I spotted the PA-28 initiating its departure on Runway 25. Recognizing the potential conflict, I quickly maneuvered my aircraft off the runway towards the grass to avoid the oncoming traffic. Fortunately, there were no close taxiways, and the grass provided a safe option. As I turned, the PA-28 aborted its takeoff and rolled down the next taxiway, allowing me to safely follow at a distance.

After completing the landing and seeing that the PA-28 had moved out of the way, I taxied and departed from Runway 07. I later learned that the other aircraft had been on a departure roll for Runway 25, but the situation was avoided without any harm.

I attribute the potential conflict to the PA-28 pilot possibly being too reliant on their radio to aid in traffic awareness and potentially missing the visual cues of my aircraft. It's a reminder that, in situations like these, visual vigilance is key - especially when operating without the aid of a radio.

Lessons Learned:

Flying without a radio demands sharp visual awareness - but it's not just my responsibility. On that day, a PA-28 began departing Runway 25 just as I was rolling out on 07. I veered into the grass to avoid a conflict, and fortunately, they aborted.

This incident reinforced that radio calls aren't enough - eyes must confirm what radios can miss. In shared airspace, especially at non-towered fields or when dealing with non-radio aircraft, all pilots must scan carefully and expect the unexpected.

Vigilance isn't optional. Visual contact saves lives when transmissions don't tell the whole story.

NOTES:

OPPOSITE DIRECTION OPS
UNSPECIFIED AGRICULTURAL AIRCRAFT & CESSNA 180 SKYWAGON

It was just another quiet afternoon near Ontario Airport (ONO), Oregon. I'd been flying aerial applications southeast of the field for most of the day. With light, variable winds, we had settled into a rhythm - departing Runway 14 and landing on 32. It had worked well up to that point.

At around 2:00 PM, I was setting up for a right base to final on Runway 32. I began the turn to short final - routine, familiar. But then, there it was: a Cessna 180, low and coming straight at me on short final for Runway 14.

We were both inbound for opposite runways.

Both of us initiated missed approaches to the right. No radio calls. No landing lights from the other aircraft. Just visual contact.

I entered a right downwind for Runway 14. The Cessna followed behind me and landed after I did. As I taxied to the loading site, the pilot of the Skywagon pulled in behind me, clearly irate.

We exchanged words.

I told him plainly, "At an uncontrolled field, turning your landing light on makes a difference - it helps others see you." He didn't like that.

But the truth is, terrain just north of ONO rises nearly 250 feet. Combine that with ground clutter, and spotting another aircraft at low altitude - especially head-on - is nearly impossible.

We were never closer than 3/4 of a mile, but that was luck - not planning.

Lessons Learned:

Uncontrolled airports demand more than procedures - they demand proactive visibility. This close call reminded me that even with clear skies and light winds, communication gaps can close distances fast.

We were lucky to spot each other in time, but luck isn't a strategy. Turning on your landing lights isn't just courteous - it's critical. Especially in areas with rising terrain and visual clutter, even broad daylight isn't enough to guarantee you'll be seen.

A radio call, a light switch, a moment of clarity - these simple actions can prevent heart-pounding near misses. At every non-towered airport, assume you're invisible until you prove otherwise.

Make yourself seen.

Make yourself heard.

Make it home.

NOTES:

RUNWAY ROULETTE
UNSPECIFIED AGRICULTURAL AIRCRAFT

It was a clear morning at Georgetown (GED), visibility about five miles and a decent ceiling overhead. I was coming in to land after an early spray mission, joining the traffic pattern for Runway 22.

Ahead of me in the pattern was a student pilot and instructor in a Cessna 152. They were flying a wide, extended downwind - wider than I'm used to seeing. I called in over UNICOM, announcing my position as number two on downwind, and kept an eye on their position. As they finally turned an unusually long base leg, I continued to tighten my own approach.

With the Cessna touching down as I was turning final, I expected them to do a typical touch-and-go. Runway 22 is 4,000 feet long and over 100 feet wide, with a midpoint taxiway 2,000 feet in. More than enough room, or so I thought.

But instead of rolling through and lifting off, the Cessna came to a slow crawl near the intersection. I was already committed and rolling out on final. I watched them closely - ready to go around - but they eventually cleared the runway just as I rolled past the midpoint. We were never closer than 1,000 feet horizontally, but it felt tighter than I liked.

In hindsight, I should have powered up and gone around when I saw they weren't moving with urgency. I let my impatience get the better of me, frustrated with the length of their pattern and their slow pace. Later, I heard they'd lodged a complaint, suggesting I shouldn't have landed while they were still on the runway.

Lessons Learned:

At uncontrolled airports, impatience is a dangerous passenger.

I assumed the Cessna would roll and go, but assumptions aren't clearance. When their extended pattern turned into a sluggish rollout, I was already committed.

I had space, but not the certainty I needed.

In hindsight, the smarter move was to go around the moment their intentions looked slow. Student pilots don't always behave predictably and neither do patterns.

My job is to adapt, not react. Next time, I'll buy myself time, not borrow it. A few extra minutes in the air beats a complaint or a collision.

In traffic patterns, clarity, margin, and courtesy are always better than threading the needle.

<u>NOTES:</u>

SHARING AIRSPACE WITH A DRONE
AIR TRACTOR AT-502

It was a sunny day with perfect visibility - ideal conditions for aerial application. I was flying an Air Tractor 502, working a field 17 nautical miles north of the Airport. As always, I began with a reconnaissance loop over the area, scouting for obstacles or anything that could interfere with my operation.

During my sweep, I spotted a truck parked a quarter-mile south of the field. On closer inspection, I realized it was the base for an unmanned spray drone operator. The drone had just landed on the truck, and given the setup, I assumed it would remain low, staying within the field's confines while it sprayed. Confident that our operations wouldn't overlap, I began my task, flying east-west passes across the field.

For the first 15 to 20 minutes, everything went smoothly. With the main application completed, it was time for the final trim passes along the field's edges - these required north-south runs. I finished the western edge, circled back, and descended for my last southbound pass on the east side of the field.

As I pulled up from the field, I spotted a drone. It was at my 2 o'clock, just 80-100 feet above me and no more than 1,000 feet away horizontally. I reacted instantly, pulling up hard to ensure I'd clear it.

Relief flooded in as I avoided a collision, but frustration quickly followed. That drone had no business being at that altitude. Spray drones, like manned agricultural aircraft, are meant to stay low over the fields they work on. This one was flying high enough to be a serious hazard, especially for pilots like me.

I had made every effort to ensure safe operations. I had circled the field at 300-400 feet AGL during my reconnaissance, clearly signaling my presence. The drone operator had to have known I was there. If there had been a reason to fly the drone so high, the operator should have waited until I left the area.

Had I approached that final pass from a slightly different angle, the outcome could have been catastrophic.

This incident underscored an urgent issue: drone operators must yield to manned aircraft, particularly in low-altitude environments. Their responsibility is to maintain separation and stay clear of manned operations.

For those of us flying low, vigilance remains our best defense. But it's clear that education and stricter enforcement for drone operators are essential to prevent close calls - or worse - from happening again.

Lessons Learned:

Low-altitude airspace is a dangerous place to assume anything - especially when manned and unmanned aircraft share the same skies. That day, I did everything right: a proper recon pass, clear pattern work, and visible flight paths. Yet, a drone still climbed unexpectedly into my flight path during a final pass.

Drone operators must understand the critical importance of yielding to manned aircraft at all times. Spray drones aren't exempt from basic airspace etiquette.

Their climb profiles, autonomous functions, and visibility limitations make them especially hazardous if operated without discipline and awareness.

This close call could have easily become a midair collision. At low level, pilots don't have the luxury of reaction time - we rely on known patterns, visual cues, and trust that others are doing their part to keep the airspace safe.

Education isn't enough. We need clear protocols, operator accountability, and, when necessary, enforcement. Until then, pilots must assume the unexpected and stay sharp. Our margin for error is razor-thin - and shrinking.

NOTES:

SPLIT-SECOND SEPARATION
UNSPECIFIED AGRICULTURAL AIRCRAFT

It was a hot summer afternoon at PBF. I lined up for takeoff on Runway 36, flying a solo ag mission under Part 91. With 13,000 hours under my belt - over 5,000 in ag aircraft - I was confident in the pattern and the field.

I had checked the surroundings and traffic, and all looked clear. But just after liftoff, something caught my eye - a bright landing light dead ahead. An aircraft was on final approach to the opposite direction, Runway 18.

At barely 200 feet AGL, I didn't have altitude on my side. I immediately rolled left to create separation, keeping my eyes locked on the approaching aircraft. It held its course. As I leveled out and tried to reassess, I realized the other aircraft had also adjusted - it was no longer on its original path but now at my 2 o'clock position.

Still holding at low altitude to avoid further conflict, I watched as the other aircraft climbed above me, finally changing its descent profile. I briefly lost sight of it but caught its silhouette crossing directly overhead. The vertical separation was tight - maybe 300 feet - and far too close for comfort.

There had been no communication, no traffic call, and no indication that the other pilot saw me until the last possible moment.

Lessons Learned:

This near miss hammered home a simple truth: visual checks aren't enough.

In uncontrolled airspace, communication is the safety net that too often goes unused. I assumed the pattern was clear based on what I saw - but I didn't broadcast my intentions. That silence nearly turned into a collision.

From now on, I'll make the call, every time. Even if I think I'm alone.

And for pilots on long final approaches, don't rely on the idea that the runway is clear - verify it. A simple radio transmission could have changed this entire encounter. Communication might feel optional in Class G, but when lives are on the line, it's a non-negotiable.

NOTES:

THE GHOST ON THE RUNWAY
AIR TRACTOR AT-301

It was a typical afternoon at Imperial (IPL) under clear skies and visual conditions. With a fresh load onboard, I began taxiing for departure. The wind sock showed a light southwesterly breeze - almost perpendicular to both Runway 14 and 32. I opted for Runway 14. It was the shortest taxi and all looked clear.

I completed a visual check - runway and airspace both appeared clear - and I called out my intentions. Confident the path was free, I taxied into position and began the takeoff roll.

As the tail came up, giving me better forward visibility, I saw it: a light-colored Cessna, either a 182, 206, or 210, already on the runway and headed straight toward me.

There was no time to hesitate.

I yanked the stick and veered hard right to exit the runway. At that moment, the Cessna lifted off and soared over me with maybe 40 feet of clearance.

I never saw it again. I don't know where it came from, where it went, or what its intentions were. No radio calls. No landing light. It blended perfectly with the background of fuel tanks, runway pavement, and sunlit haze. A ghost with wings.

Looking back, if that pilot had simply turned on their landing light, I might have spotted them sooner. If I'd waited another moment before taking the runway, this report might have been a very different one.

Lessons Learned:

This encounter reminded me that the biggest threats can come from what you don't see - or hear.

At uncontrolled airports, it's easy to trust that a clear radio and a visual scan mean you're in the clear. But assumptions are dangerous. That Cessna was invisible until it nearly became fatal.

From now on, I'll wait an extra beat. I'll scan twice, maybe three times. I'll expect the unexpected. And I'll always use every available tool - lights, calls, strobes - to make myself seen and heard.

Familiar fields don't guarantee safe departures. Each takeoff demands full attention, not just to what's visible, but to what might be hiding in plain sight.

NOTES:

THE SILENCE BETWEEN TWO WINGS

GRUMMAN G-164B AG CAT & G-164D AG CAT

It was a clear May morning in Louisiana. Hot already, the sun pressing down as we pushed through another long day of ag work. I was flying the B model. My buddy was in the D. Same strip, same routine - we'd done this countless times before. No radio chatter, just rhythm. One bird landing, the other loading. In and out like clockwork.

I was in the air, wrapping up a pass, and I saw him at the tender truck. He was getting topped off, not quite on the runway yet. I figured I had time. I turned final, set up for landing in the usual direction, and committed.

Wheels down. Rollout smooth. The dust was just beginning to rise behind me when I saw something out of the corner of my eye - movement ahead. Fast. Closing.

It was him.

He'd finished loading, taxied into position, and started his takeoff roll - straight at me. Same strip. Opposite direction.

In a flash, I tried to veer, pulling the stick, pushing rudder, hoping I could duck or dodge. But there was no space. No time. His left wing collided with mine.

The sound of aluminum tearing through aluminum echoed across the field, a sickening crunch that drowned out the roar of engines.

We came to a rest, both of us stunned. The wings on both aircraft were mangled, our fuselages bent and battered. Stabilizers, elevators - smashed. But we were both okay. No injuries. Just a pair of lucky, shaken pilots stepping out of wrecked machines.

Later, when we picked through the damage and ran through the events, it was clear: neither of us had made a radio call. No position reports. No runway announcements. Nothing.

We were operating under the assumption that we'd see each other. That our timing would sync. That routine would keep us safe.

But routine is a liar. It tells you everything will be like the last time. Until it isn't.

Lessons Learned:

Communication isn't optional - it's essential.

That morning, we were two professionals with nearly 40,000 combined hours between us. But all that experience didn't count for much when we failed to speak.

In ag operations, especially at private strips, it's easy to fall into habits. You start trusting patterns instead of procedures. You think, "He'll wait." Or, "I'll be quick." You assume. And assumption is the most dangerous co-pilot in the cockpit.

If either of us had keyed a mic and said, "Taking off southbound" or "Landing northbound," this story wouldn't exist. We'd have passed each other by with a wave, not a wing strike.

Now, I don't care if the strip's remote, if I think no one's around, or if I've flown it a hundred times - I always announce. Every time. Because I've learned that silence between two wings can echo a lifetime if you're lucky - and end a life if you're not.

NOTES:

THE SILENT RUNWAY CONFLICT
UNSPECIFIED AGRICULTURAL AIRCRAFT

It was a typical afternoon in Brawley, California. I had been running spray loads all day in an ag aircraft without a radio. The wind was steady out of the southeast, and I was using Runway 08.

After picking up another load, I taxied out and stopped at the hold-short line. I scanned the approach paths for both ends of the runway - clear. Confident, I pulled onto the runway, lined up on centerline, set my flaps and condition lever, locked the tailwheel, and began my takeoff roll.

Just a few hundred feet into the roll, a cloud of dust rising from the north side of the east end of the runway caught my eye. Then I saw it - a light tan Cessna 152, barely visible against the desert backdrop, sitting in the dirt near the far taxiway.

The Cessna had either aborted a takeoff or landing, but one thing was clear - we had both used the same runway, at the same time, in opposite directions. By the time I spotted the aircraft, I was nearly at rotation speed. I lifted off, sidestepped to the right (south), and climbed out, passing safely away from the other aircraft.

Looking back, the Cessna must have entered the runway while I was configuring the aircraft for takeoff.

Or I simply failed to spot him on final.

His beige color blended in perfectly with the dusty surroundings, and with the sun low behind me, his visibility may have been just as compromised.

But one factor loomed largest: no radio.

Lessons Learned:

In uncontrolled airspace, especially when operating from remote or dusty strips like BWC, visibility and communication are critical. I had no radio. The other pilot might not have seen me. Neither of us was able to confirm intentions, positions, or timing. The result? A near head-on conflict on the runway that could have been catastrophic.

Color schemes that blend into the terrain, low sun angles, and a lack of radio communication can stack the odds against safe separation. The solution isn't complex: when flying without a radio, increase your vigilance tenfold. Taxi slower. Delay takeoff a few more seconds. Scan again - and again. And if a radio is an option, make it a priority.

That day, luck and timing kept two airplanes from colliding. But luck isn't a safety strategy. Awareness is.

NOTES:

UNANNOUNCED APPROACHES
UNSPECIFIED AIR TRACTOR & UNSPECIFIED AIRCRAFT

The afternoon arrival into PRN Airport should have been routine. The weather was clear, the winds were favorable for Runway 14, and our flight was proceeding smoothly under visual flight rules. My first officer and I were operating a light transport aircraft under Part 135, carrying passengers. As we descended, we monitored both Atlanta Center and PRN's CTAF frequency, making standard position reports. The radio was eerily silent.

With a clear visual of the airport, we canceled IFR but remained on with Atlanta Center for traffic advisories. Several targets still appeared on our multifunction display (MFD), though Center did not show them. We switched over to CTAF and announced our planned entry onto a 45-degree downwind for Runway 14. Still, no response from other aircraft.

Approaching pattern altitude, we heard a single transmission from an Air Tractor pilot stating he was departing the area to the southwest. We identified his position on the MFD, approximately 1,700 feet below us. With no further radio traffic, we proceeded with a normal approach and landed without issue.

As we rolled out and turned toward the last available taxiway, something caught my copilot's eye. An aircraft was approaching us head-on, a mere 100 feet above the approach end of Runway 32 - clearly setting up to land in the opposite direction.

We had not yet called clear of the runway, still slowing as we reached our taxiway turnoff. The approaching aircraft, another Air Tractor, had not made a single radio call. My copilot yelled for me to clear the runway. Without hesitation, I made a sharp right turn onto the taxiway as the Air Tractor landed, rolling past us as we continued toward the ramp. The proximity was uncomfortably close.

Once shut down, we observed the Air Tractor taxi to the east ramp area and, moments later, depart from Runway 14. Then, another Air Tractor appeared - making a short approach at a low level, banking sharply to land on Runway 32, before taking off from 14 again. No pattern, no calls. The arrivals and departures were unpredictable, emerging from behind the tree line at low altitude, descending rapidly onto the runway with steep angles of bank. It became apparent that these crop dusters were running their own operation, completely separate from standard airport procedures.

Even after our passengers disembarked, we continued monitoring CTAF and watching the activity. Still, not a single radio call from the Air Tractors. After some inquiries, we learned these aircraft had just arrived and were conducting aerial spraying operations. Despite this, there were no NOTAMs indicating such activity.

Lessons Learned:

- Maintain Vigilance Beyond CTAF Calls – Even if you don't hear other aircraft, that doesn't mean they're not there. Use all available tools, including onboard traffic displays.
- Expect the Unexpected at Uncontrolled Fields – Agricultural aircraft may operate outside standard traffic patterns and procedures, particularly at remote airports.

- Exercise Caution When Canceling IFR – If traffic is unknown or unclear, staying with ATC for advisories can provide an extra layer of safety.
- Be Ready to Take Immediate Action – Situational awareness and quick reactions prevented a dangerous conflict on the runway.

Uncontrolled airports come with their own unique risks.

In this case, we avoided an incident, but had we been unaware, the results could have been catastrophic.

Always expect the unexpected - especially when silence fills the airwaves.

NOTES:

UNEXPECTED DRONE ENCOUNTER
AIR TRACTOR AT-602

As the pilot-in-command of an Air Tractor 602, I was hauling a load of pesticides to a field early that morning. The skies were clear, and visibility was perfect - ideal conditions for the task ahead. Upon reaching my fields, I noticed a drone and its tender truck operating just under a quarter-mile away, spraying a field.

I decided to fly a parallel path to the drone, working east-west on my first field. As I continued my passes, the drone stayed in its pattern, spraying the field below. After roughly 12 minutes of working the first field, I finished and moved to the next, which had me flying the same east-west route, but with my turns required over the ongoing drone operation.

Before making my first pass over the drone, I circled and puffed smoke to get the operator's attention. They continued their operation without any sign of acknowledgment. I began my application runs, making about 25 passes over the drone's location, with each turn at a low altitude, putting me roughly 100 feet above the drone.

Despite this, the drone operator did not adjust their course or respond to my smoke signals. They kept spraying the field, landing, reloading, and taking off again as I continued my work.

The operator appeared to be on the bottom level of a multi-level trailer, likely unable to see their drone while it was airborne.

The lack of communication and situational awareness on their part put both our operations at risk.

I've flown in close proximity to other aircraft before, but this encounter with the drone operator was a close call. If I had been in a slightly different position or taken a different route, the situation could have turned out much worse.

Lessons Learned:

Operating in shared airspace demands more than just staying in your lane - it requires full awareness of who else is in the sky. That morning, I was flying manned, low-level spray passes while a drone operated nearby with no visible effort to communicate, coordinate, or even maintain a line of sight.

Despite my smoke signals and multiple low passes overhead, the drone continued flying autonomously, its operator seemingly unaware or unreachable. That's not just poor protocol - it's dangerous.

Drone operators must understand that their presence in ag airspace isn't invisible or harmless. A midair collision, even with a small UAV, could be catastrophic for a manned aircraft flying at speed and low altitude. There's no time for evasive maneuvers when every foot counts.

Shared skies mean shared responsibility. That includes maintaining visual contact, using spotters, and being proactive when a manned aircraft is nearby. Communication and situational awareness aren't optional - they're the difference between a successful day's work and a headline.

NOTES:

UNSEEN AND UNANNOUNCED
UNSPECIFIED AGRICULTURAL TURBOPROP

After completing an aerial application over a farm field located approximately 10 miles southeast of LRJ airport, I was climbing out to my typical ferrying altitude of 500 feet AGL to return to LRJ. The sky was clear, and the conditions were ideal for the flight.

While climbing at 350 feet AGL, about a mile to the northwest of my target location, I noticed a white object briefly appear in the lower right corner of my windshield. At first, I thought it was a seagull, as birds are rarely white during this time of year. However, as the object reappeared under my right wingtip, I quickly realized that it was a white-colored quadcopter-type UAV. It missed my right wingtip by an estimated 50 feet horizontally and 20 feet vertically.

The encounter happened too quickly for me to take any evasive actions. The UAV passed so close that I could not avoid it in time. After the near miss, I circled the field to assess the situation. I observed that the UAV had entered a stationary hover about 350 feet AGL above the field. Upon further inspection, I saw what appeared to be the operators inside a pickup truck parked adjacent to the field, facing west. There were no visible operators or observers outside the vehicle.

I watched as the UAV began descending to the road, and the truck moved to intercept it. After observing this, I continued my flight direct to LRJ without any further issues or incidents.

Lessons Learned:

This close encounter underscored a growing concern in modern aviation - UAVs operating in shared airspace without adequate situational awareness or coordination.

As manned aircraft pilots, especially in low-altitude operations like aerial application, we expect the airspace to be clear unless notified otherwise. But drones don't announce themselves on the radio, and they're small and difficult to spot until it's almost too late

The key takeaway here is twofold: first, UAV operators must take greater responsibility to ensure they yield to manned aircraft and avoid active airspace during agricultural or low-level operations.

Second, we as pilots must maintain heightened vigilance, even in familiar areas, and anticipate the unexpected.

NOTES:

PART 5 - COMPLACENCY & FATIGUE

"A superior pilot uses his superior judgment to avoid situations that would require the use of his superior skills."
–Mike Kloch

A TURN TOO TIGHT
AIR TRACTOR AT-402

It was a clear afternoon - dry, bright, and exactly what you'd expect in the Texas Panhandle. I climbed into the seat of my Air Tractor AT-402 and got ready for another day in the field. At 23, I was already deeply into the world of agricultural flying. This work - low-level, fast-paced, and precise - had become part of who I was.

The aircraft I flew that day was no stranger to the job. Built in 1990, it had more than 10,000 hours on its frame. A tailwheel turboprop designed to do exactly what I was doing - fly low, spray chemicals with accuracy, and handle the punishing rhythm of ag ops.

The assignment that day was straightforward: liquid application over a grass field under Part 137. I launched just before midday, made several solid passes, and by early afternoon, the hopper was empty. Job complete. Or so I thought.

But in this business, the danger doesn't always come during the run. Sometimes, it shows up in what you do after it's over.

I entered a right-hand turn to head back. A standard maneuver - something I'd done hundreds of times. But that day, for reasons I still replay, I tightened the turn just a bit more than usual.

Maybe I was being efficient. Maybe I wasn't fully aware of how much energy I'd loaded onto the wing.

I was still flying low, just above the wheat, and in that steepening arc, the aircraft's angle of attack crept higher and higher. Then it passed a line - not one you can see, but one every pilot knows.

I'd exceeded the wing's critical angle of attack.

The stall was immediate. One moment I was flying. The next, I wasn't. The airplane quit. Not because of an engine failure. Not because of weather. But because I'd asked the wing to do more than it physically could. The aircraft dropped sharply, nose down. I had no altitude to work with.

We hit the field hard. And that's when the second threat arrived.

A post-crash fire erupted instantly. The impact had ruptured fuel lines, and dry conditions did the rest. Flames swept through the aircraft with terrifying speed.

Inside, heat and smoke pressed in. I was disoriented, hurt, but conscious. I unbuckled, fought through the rising panic, and forced my way out. Somehow, I escaped.

I was seriously injured, but alive.

Behind me, the Air Tractor was gone. Flames had consumed nearly everything - only scorched remains and twisted metal were left. The same aircraft that had just been part of my daily routine was now a skeleton in the grass.

Emergency crews arrived quickly. They secured the scene, but there was little left to save.

The investigators found no mechanical fault. The engine and flight controls had been functioning normally. The weather was clear. No wind shear, no turbulence.

What happened wasn't a failure of the aircraft. It was a stall. A simple, devastating aerodynamic stall.

I'd exceeded the aircraft's limits in a low-level turn and lost control. There was no altitude left to recover. The cause wasn't hidden in the engine or buried in maintenance logs.

It was in the decision I made in that moment of habit - tightening the turn too much, too low.

Lessons Learned:

Stalls aren't just theory. They're real. They're deadly. And near the ground, they give you no time at all. In ag flying, where precision is everything and time feels compressed, it's easy to normalize aggressive maneuvering.

But physics doesn't care how many hours you've logged. It doesn't care how many good passes you've made. The wing has limits. Exceed them, and flight turns to free fall.

This didn't happen during the mission - it happened after. A final turn. A moment of routine. And it nearly ended everything.

Respect the wing. Respect the margin. Fly the airplane like your life depends on it - because it does.

The job's not done when the tank's empty. It's done when you're back on the ground, safely. And that last turn? That's the one that counts the most.

NOTES:

COMPLACENCY AT LOW ALTITUDE
GRUMMAN AG-CAT TURBO

During a routine agricultural flight spraying fungicide on a cornfield, I completed my final pass parallel to a power line running through the field and began turning to apply a 'clean-up' pass.

As I prepared for the 90-degree turn, I became focused on the area around a barn and house, scanning for possible drift-sensitive areas and bystanders. In doing so, I lost situational awareness of the power line directly ahead and flew through it. The collision resulted in the breaking of several cables ranging from 1/2 to 5/8 inch diameter.

Although the aircraft suffered only minimal structural damage, the event was serious and could have had far more severe consequences. I attributed this lapse in attention to a combination of factors. Despite feeling well-rested after not flying the day before, I had logged 45 to 60 hours of spraying over the prior six days. The physical toll of long hours, combined with short nights of sleep, left me feeling fatigued, and I had not fully recovered. In addition, the pressure to catch up on work led to self-imposed stress, which in turn affected my focus.

This is a common hazard in the agricultural flying industry, and although the aircraft was designed to withstand wire strikes, the event served as a stark reminder of the dangers present in low-altitude, high-threat environments.

Lessons Learned:

This event is a clear example of a human factors issue. It emphasizes the importance of maintaining situational awareness, even in familiar or routine operations. The distractions caused by focusing on peripheral issues like buildings, people, and potential drift-sensitive areas caused me to overlook the immediate danger posed by the power line. Despite my extensive experience, including years of combat jet fighter flying, I became complacent and allowed my attention to wander.

The key takeaway is that fatigue, self-imposed pressure, and rushing to "catch up" can significantly impair decision-making and focus. No matter how experienced a pilot is, it's essential to prioritize safety over everything else, even in the face of tight schedules. Adequate rest, proper hydration, and maintaining focus on primary tasks are critical to avoiding accidents.

In a high-risk industry like agricultural aviation, human error can have dangerous consequences, and this event serves as a reminder that attention to detail and situational awareness are paramount.

NOTES:

DARK DESCENT
BELL OH-58A

It was a black, windless night in the desert - ideal conditions for one last rinse flight. I'd already completed a chemical application run and just needed to flush the system. I climbed into the cockpit of my old Bell OH-58A and lifted off with 50 gallons of clean water in the tank. The rinse was routine. I'd done this dozens of times.

The field I was headed to was just a couple hundred yards away - barely a hop from the load truck. I lifted off and pointed toward it, staying low, around 40 feet above the ground. That's when things went wrong.

The helicopter started behaving oddly. It was twitchy - fishtailing, bobbing, yawing in all directions. It felt unstable, like I was on the edge of losing control. I adjusted the stick, worked the collective, trying to smooth it out. There was no visibility - just blackness in every direction. I couldn't judge the terrain. I was flying blind.

Then came the silence.

The engine quit. Completely.

I focused on leveling the aircraft. Pulled back, managed the descent as best I could, but I had nothing visual to work with.

It was just instinct and feel. No power. No lights. Just black air and fast-approaching dirt.

I hit the ground hard.

The main rotor chewed through the tail rotor driveshaft. The airframe jolted. The impact was violent. A moment later, the entire main rotor and blade assembly tore off and landed far from the wreck.

I sat still, shocked, surrounded by bent metal and silence again - but this time with the hum of disbelief. I had survived. Somehow. A few bruises, nothing major. But the helicopter was wrecked.

After the accident, investigators combed through everything. There was no fire, no structural failure, nothing visibly wrong. The controls had worked. The fuel pump worked. The engine showed no signs of mechanical failure. Throttle linkages were intact. They even bench-tested the engine - no issues found.

There was fuel in the system - five gallons in the tank. But only 1.4 of that was usable.

The gauge? When they powered it up, it showed nearly empty. Not quite zero - but close. It's possible I saw that and assumed there was still enough to get through the rinse. Maybe I trusted my estimates over the instrument. Maybe I got complacent. Either way, the margin was razor-thin.

Despite all the data, the final verdict was frustratingly vague: complete loss of engine power - reason undetermined. No single mechanical fault. No smoking gun. Just a clean-running aircraft that suddenly lost power at the worst possible time.

The Bell had been maintained meticulously. The engine had been replaced just a few months earlier. A fresh 100-hour inspection had been done. Everything had looked solid. But that night, something tipped the balance. A nearly unusable fuel level. A bad gauge. A night op with no visual cues. It was a stack of small risks that added up to one big fall.

Lessons Learned:

This flight taught me something every pilot should remember: not all failures come with clear reasons - but the risks often show up early if you're paying attention.

Here's what I took from it:

- Don't rely on faulty gauges. If it reads low, believe it - or better yet, verify it yourself.
- Usable fuel matters. It's not about what's in the tank - it's about what gets to the engine.
- Never treat rinse flights as casual. They still involve flight. Every takeoff has stakes.
- Flying at night with no visual reference will multiply any existing problem.
- When you're down to fumes, you're already out of options.

I didn't crash because of one decision - I crashed because of a series of small, unchecked risks. The kind we all think we can manage - until we can't.

This one ended with a broken aircraft and a survivor. But I know how easily it could've gone the other way. The lesson? Never let the end of a long day be the start of cutting corners.

NOTES:

FAST TAXI, HARD LESSON
GRUMMAN AG-CAT TURBO

On what seemed like a normal day, I was conducting a maintenance check by taxiing up and down the runway at a high speed. The visibility was perfect, the weather was ideal, and the conditions were as expected for a typical test flight. The plan was simple - just to assess the maintenance of the aircraft by taxiing it on the private grass strip I own, with no intention to fly.

As I approached high speed during the taxi, I suspect I may have hit a badger hole, although I'm not entirely sure. The runway is not level, which posed a potential hazard. The moment I hit the rough spot, the aircraft ballooned to the right, causing the tire to fly off. The loss of the tire led to a loss of control, and the aircraft accelerated uncontrollably before flipping over onto its top.

Fortunately, I was able to exit the plane without injury. The aircraft sustained major damage, but there was no other property damage, as the incident occurred on my own land. The visibility, meteorological conditions, and my performance all seemed normal, and I was confident in my taxiing abilities. There was no indication that anything would go wrong, and my judgment and decision-making remained sound.

Looking back, I realize that this incident could have been avoided with a more careful inspection of the runway conditions and a more cautious approach to taxiing at high speed, even when no flight is intended.

Lessons Learned:

Maintenance checks may seem routine, especially on private property, but they demand the same level of care and caution as any flight operation.

High-speed taxiing on uneven or uninspected surfaces introduces serious risk - even without liftoff. In my case, a hidden hazard like a badger hole likely triggered a chain of events that flipped the aircraft.

The lesson is clear: always treat high-speed ground operations with respect. Thoroughly inspect your runway or taxi area beforehand, no matter how familiar it seems. Conditions can change without warning. What appears safe may hold unseen dangers.

Vigilance on the ground is just as vital as it is in the air.

NOTES:

FUEL FOR THOUGHT
AIR TRACTOR AT-301

The skies were calm and clear that morning as I cruised above the Missouri landscape in my Air Tractor AT-301. I was en route between Lamoni and a spray site near Andover, with 1,700 pounds of dry fertilizer in the hopper and a smooth-running radial engine up front. The kind of morning you hope for - no wind, no turbulence, just smooth air and a job to do.

At 23, I had close to 1,000 flight hours, most of those as pilot in command. I'd grown confident in the seat of that aircraft. This wasn't new, and nothing in the first few minutes of flight suggested anything out of the ordinary.

Then I noticed the aircraft wasn't holding altitude. Subtle at first, just a slight sink rate. I adjusted pitch, checked my airspeed. It continued to drop. Something was wrong. My hands moved automatically: throttle full forward, mixture rich. I scanned the gauges - RPM, manifold pressure, oil temp, fuel pressure - everything in the green. No warnings. No alerts.

But the airplane kept descending.

The engine didn't sound rough. In fact, it was smooth, steady, almost deceiving. But it wasn't pulling.

I was light on the stick, coaxing performance that just wasn't there. With terrain and trees approaching, I didn't have time to wonder. I dumped the load.

Pulling the jettison handle, I felt the release as 1,700 pounds of fertilizer fell away. The aircraft lightened immediately, but not enough. The sink rate continued. I started scanning for a field - somewhere to put it down. I found a spot ahead and committed.

As I descended, the left wing clipped a tree. It yanked the aircraft sideways, and I didn't have the altitude or control authority to recover. We hit hard. The gear slammed into the field, the frame absorbing the impact with a force that rattled through my bones.

But we stayed upright.

The engine tore free from its mount. The wings crumpled. But there was no fire, no fuel leak, no explosion. I sat there for a moment, strapped in, stunned. Then I opened the canopy, stepped out, and looked back at what was left. Bent metal in a quiet Missouri field.

The investigation moved quickly. With no fire damage, it didn't take long to piece together the puzzle. The tanks were full, but something was off. The fuel smelled wrong - almost sweet. Its color wasn't what it should've been. The samples were sent for testing.

It came back as regular 87-octane automobile gas. Not aviation fuel. Not STC-approved mogas. Just straight pump fuel - probably with ethanol. Bought from a local gas station. Not only was the aircraft not certified for it, but the owner hadn't checked for ethanol content, hadn't mixed it, hadn't verified anything. Just filled it up and sent it out.

The radial engine I was flying behind was never designed for automotive fuel. Ethanol can be lethal to aircraft systems - eating seals, attracting water, and increasing the risk of vapor lock or carb ice. Worse, it robs power. You don't always get an engine failure - you just don't get the performance you're counting on.

And that's exactly what happened. The engine didn't quit. It didn't cough or sputter. It just didn't give me what I needed. It gave me smooth noise, and not enough thrust.

The findings were clear: the use of unauthorized fuel had reduced engine output, making it impossible to maintain altitude during cruise. That led to a forced landing, a collision with a tree, and serious airframe damage.

I did everything I could. I dumped the load. Picked a field. Controlled the descent as best I could. And I walked away.

But that airplane was never going to make it once it left the ground with the wrong fuel in the tanks.

Lessons Learned:

It wasn't a birdstrike, or weather, or a mechanical fault. It was fuel. Just fuel. And a dangerous assumption.

Flying doesn't begin with takeoff - it begins with preparation. And that includes what you pump into your wings. The wrong choice may not show itself until you need full power and find there's none left.

Always check. Always ask. Because when you're low, slow, and heavy, even a little performance loss can be the difference between flying and falling.

I flew well that day. But no amount of skill can overcome what the engine can't give. Know what's in your tanks - your life depends on it.

NOTES:

FUEL TO THE FINISH
THRUSH S2R

After 30 years in the ag flying business and more than 6,000 hours in type, I thought I'd seen it all. But on a humid July evening in Minnesota, I added a new experience to my logbook - one I never expected: running out of fuel.

It started with a straightforward spray job - 160 acres of potatoes located 20 miles from the Moorhead airport. Due to recent heavy rains, the field had been hit hard. I was told there were only 65 acres left worth spraying. Normally, you'd find those remaining patches concentrated in one section. But this field? It was chaos. The surviving potatoes were scattered throughout the entire area.

That meant I had to treat the full 160 acres, toggling the spray on and off constantly, zigzagging through wet pockets, and burning more time - and fuel - than I had planned for. By the time I wrapped up, I was behind schedule and low on fuel, but the gauges still showed some left in both wing tanks.

I turned for home and was about a mile northeast of the airport when both engines went quiet.

No panic - just instinct. I spotted a nearby bean field, lined up, touched down smoothly and rolled to a stop. Undamaged and intact.

After a walk to find some help, I returned, fueled up the bird, and flew it out without a hitch.

Of course, my competitor happened to be spraying a field nearby and watched the whole thing unfold. He called the sheriff when he saw me walk out - probably thinking something worse had happened. But there was no damage, no drama - just a hard-earned reminder.

Lessons Learned:

After 8,000 flight hours and decades in the cockpit, it's easy to let your experience convince you that you've got things under control. But experience can't replace vigilance.

Fuel gauges can be misleading. Margins can shrink quickly. And assumptions - like expecting to only spray 65 acres - can get you in trouble.

The day taught me that even after 30 years, complacency is just one decision away from consequence. Always fuel with margin. Always fly with a backup plan. And never let routine dull your respect for the unexpected.

<u>NOTES:</u>

GROUNDLOOP AT THE HOME STRIP

AERO COMMANDER S2R

It had been a solid hour in the air - another agricultural spray job over familiar Missouri farmland. I was flying the Aero Commander S2R, a tough bird, built for this kind of work. Everything had gone smoothly, and I was heading back to home base to land on runway 14. Winds were gusty - coming out of the north - but manageable. I'd flown in worse.

As I turned onto final, I felt calm. Confident. The kind of confidence that builds after you've done a job a hundred times and know every inch of the airstrip. There was a slight quartering tailwind, but nothing I hadn't handled before.

I flared over the numbers and the wheels met the concrete - but not quite right. The plane bounced. Not hard, but enough to throw things off. I instinctively worked the controls, trying to recover, correct for the wind, bring it back to centerline. I came down again, but this time the crosswind caught me. The tail swung left - fast.

And just like that, I was in a ground loop.

The aircraft veered off the runway. I fought for directional control, but it was gone. We skidded sideways off the strip, the tires digging into the grass, the plane twisting around in a sickening arc.

When we stopped, the damage was clear - bent wings, a battered empennage, and a bruised ego. I shut everything down and climbed out, thankful I was uninjured but furious with myself.

There were no mechanical problems. No unexpected malfunctions. The aircraft had been in great condition. The engine ran flawlessly. The issue was purely mine - a loss of control in gusty conditions, landing with a tailwind I should've respected more.

Lessons Learned:

It's not the wind that gets you - it's underestimating what it can do.

A six-knot tailwind gusting to fifteen doesn't sound dramatic on paper. But when it's pushing at just the wrong moment, in just the wrong direction, especially with a tailwheel aircraft, it can flip your day upside down - literally. Add in a bounced landing and a bit of complacency, and the margin for recovery disappears fast.

Since that day, I've become far more cautious about landing decisions. I'll switch runways. I'll wait out gusts. I'll go around rather than try to force a landing that doesn't feel right. Because every approach might seem routine - until it isn't.

In ag flying, we live close to the edge - low, heavy, and hot. But it's often the end of the flight, not the middle, where things go wrong. And when they do, it's the small decisions that make all the difference.

<u>NOTES:</u>

HEAVY AIR, HARD TRUTHS
UNSPECIFIED AIR TRACTOR

It was already 96 degrees by the time I climbed into the Air Tractor that morning. The skies were clear over Helm, Mississippi, but the humidity was thick. I've flown in worse-looking weather, but I know days like this can be deceptive. There's no lightning, no turbulence, no gusts - just hot, heavy air that drains lift without warning.

I'd flown more than 1,400 hours, most of them in this very make and model. I was working out of a private 2,200-foot strip, half concrete, half grass, and had already completed a few runs earlier. This one felt like more of the same. The aircraft was loaded with 350 gallons of fertilizer - within spec, but right up against the edge given the temperature.

I felt confident. The wind was calm. Visibility was perfect. The airplane had handled well on earlier flights. I advanced the throttle and rolled down the strip.

At first, everything sounded right. The engine roared and the tail came up. But as I reached the halfway mark on the runway, something didn't feel right. We weren't climbing. We weren't even accelerating like we should've been. The power was there - but the lift wasn't.

That's when it hit me: density altitude.

Even though I was flying out of a field just 125 feet above sea level, the aircraft was behaving like it was already at 2,500 feet. The heat and humidity had changed the equation. The engine was making noise, but the wings were struggling to do their job.

By the time I realized what was happening, I was near the end of the strip. Trees loomed ahead. I reached for the dump lever and released the load - 350 gallons gone in an instant. But it wasn't enough. Not soon enough.

The airplane stalled just past the runway's edge. The nose dropped, and the wings - lighter now but still fighting thick air - couldn't hold lift. I hit hard, just beyond the field.

The impact was brutal. The wings folded. The fuselage crumpled. The engine bay buckled, and the empennage twisted. But somehow, there was no fire. The aircraft came to a stop upright, and my harness had done its job. I was shaken, but not injured. I crawled out of the wreckage into the sweltering air, surrounded by the broken remains of a machine that had done everything it could.

The investigation followed quickly. I didn't hold anything back. I told them I'd flown multiple runs that morning. I knew it had gotten hotter. And yes - I admitted it. I should've cut the load. It made sense earlier in the day, but by that last flight, conditions had changed. I hadn't changed with them.

They found no mechanical fault. The engine was fine. The controls were fine. The problem wasn't with the airplane - it was with the air. The aircraft had done exactly what physics dictated. The density altitude had quietly robbed me of the performance I needed, and my response came too late.

The final report was clear: loss of control on takeoff due to an aerodynamic stall. Contributing factors: high density altitude and a load that was too much for the conditions.

A hard lesson, but an important one. Hot, heavy air doesn't look like a problem. It doesn't rattle your headset or blow dust across the field. It just sits there - silent, still - and waits for you to assume things are normal.

Lessons Learned:

You can't see density altitude, but you'll feel it the moment you expect performance and it doesn't come. It doesn't care how many loads you've flown. It doesn't care how confident you are in your machine. It just changes the rules, silently, and waits for you to find out too late.

I survived because of training, instincts, and luck. The airplane - strong and well-built - took the hit and protected me. But the wreckage told the real story.

So here's the takeaway: fly lighter when the air is heavy. Do the math. Don't trust the day to stay the same. Give yourself margin, because even a perfect airplane can't fly in impossible air.

And when in doubt - leave a little behind so you can fly again tomorrow.

NOTES:

RUNAWAY MISTAKE
UNSPECIFIED AGRICULTURAL LOW WING AIRCRAFT

It was a cold Tennessee morning at APT. I'd just started up the aircraft for an agricultural flight - routine stuff. After checking the oil and draining the fuel sumps, I climbed in and got the engine turning. With the plane idling smoothly at around 600–700 RPM and parked off the pavement in a grassy area sloping slightly away from the ramp, I figured I was safe.

I untied the aircraft, expecting it to stay put. The ground was firm, the slope gentle, and the aircraft heavy. What could go wrong?

With the engine still running at low RPM, I turned to finish the rest of my preflight, not expecting anything out of the ordinary. But within 30 seconds to a minute, the unexpected happened.

The aircraft was rolling.

By the time I turned around, it was already too late. The plane had drifted across the grass and right into another parked aircraft. The result: a crumpled left aileron and a damaged wingtip. All from what I assumed was a safe, stationary setup.

I'd overlooked a key factor - the wind. On that chilly morning, a breeze had picked up just enough to nudge my immovable aircraft.

And in the absence of wheel chocks or close attention, gravity and momentum took care of the rest.

Lessons Learned:

Even on calm days, a slight slope, light wind, and an idling engine can turn into a costly mistake.

I assumed the aircraft would stay put - untied, unchocked, and idling at low RPM. Within seconds, it had rolled across the grass and into another plane.

Damage was immediate and preventable.

From now on, I chock the wheels before engine start, stay with the aircraft at all times while running, and never underestimate environmental factors.

Safety on the ground is just as critical as in the air. Because one moment of inattention can cause a wreck without ever leaving the ramp.

NOTES:

SETTLING INTO TROUBLE
UNSPECIFIED AGRICULTURAL HELICOPTER

It was a calm morning in Idaho, and I was wrapping up the last load of the day - just one more pass to finish a spot application on some stubborn Canadian thistle. With only half fuel on board and ideal VMC conditions, I initiated a standard agricultural turn, a maneuver I'd executed countless times in my 2,800 flight hours.

But this time, the air had other plans.

Mid-turn, I was hit by a sudden gust of tailwind - a textbook wind shear. Instantly, I felt a sharp drop in altitude, and the unmistakable onset of a settling-with-power scenario. Rotor RPM dipped. Manifold pressure was sluggish. My helicopter was losing lift, fast. I kept the nose down and tried to power through into cleaner air, but it wasn't working.

Realizing how quickly I was approaching the ground, I dumped the load, pulled pitch, and rotated to arrest the descent - but I hit the ground hard on rising terrain. The gear collapsed, rotor blades struck the airframe, and the aircraft twisted to a stop, 90 degrees off heading. I scrambled out, only to see fuel spilling onto the hot turbo.

The fire started immediately.

I used my handheld halon extinguisher, but it was ineffective.

Thankfully, a nearby water truck helped suppress the flames enough to prevent further spread, and the aircraft was eventually removed from the field.

Looking back, the warning signs weren't obvious. This was a textbook gray zone on the height/velocity chart - a helicopter's Achilles' heel. I was operating at low airspeed, low power, and got caught at the top of the turn, the moment of greatest vulnerability. In hindsight, I should have initiated recovery sooner or jettisoned the load earlier. Most critically, I should have opted for a different style of turn given the terrain and potential for wind shifts.

The day had started perfectly - low stress, good rest, and a routine job - but this incident was a stark reminder: even when everything feels right, conditions can change in a heartbeat. Always expect the unexpected.

Lessons Learned:

In helicopter ops, smooth conditions can lull you into false security. But the top of a turn remains one of the most vulnerable phases of flight.

That's where I got caught: low airspeed, rising terrain, and a sudden gust that pushed me into settling with power. I was outside my escape window before I knew it.

The lesson? Respect the height/velocity diagram like gospel. Dump the load early. Fly the turn with enough margin to handle the unexpected. And remember, calm skies don't mean stable air.

Agricultural work demands precision, but never at the cost of recovery options. Plan for the gust, expect the terrain shift, and always fly with an exit in mind.

NOTES:

THE BERM I DIDN'T SEE
AIR TRACTOR AT-602

The heat was already rising as I prepped for my third load of fertilizer that day. The job was going smoothly - standard application work over open fields out of the turf strip at Newport. I was flying the Air Tractor AT-602, a machine I knew well. We'd handled heavier loads and shorter strips before. With three-quarters fuel in each tank and a full hopper, I lined up for takeoff, confident this would be another routine run.

I pushed the throttle forward, the engine roaring to life as we tore down the grass runway. The aircraft began to lift, straining with the weight but holding steady. I was just one to two feet off the ground when it happened.

The airplane settled back down.

In that instant, I felt it - an impact on the left side, a jarring lurch as the left main landing gear hit something solid. The plane bounced slightly, but the gear was gone. I had no choice - I needed to get airborne again or risk a total loss. I reached for the hopper dump, released the fertilizer load, and pulled the aircraft into a gentle turn, circling back toward the strip.

I knew the landing would be tricky.

With the left main gear missing, I'd have to balance the aircraft on the right main and tailwheel, holding the left wing up as long as I could. I touched down softly, the right side rolling smoothly while I eased the left wing lower, trying to delay the inevitable.

The moment the left side touched, I felt the structure give way. The wing dug into the turf, dragging hard. The aircraft skidded sideways and came to a stop in a mess of bent metal and crumpled gear. I was uninjured. Shaken, but okay.

Later, standing at the edge of the strip, I found the culprit.

A berm. Completely hidden by tall grass at the end of the runway. I hadn't seen it on the earlier takeoffs. On this one, the aircraft hadn't cleared it and it tore the gear off.

There were no mechanical issues. The plane had been in great shape. The takeoff conditions seemed suitable, but in hindsight, I'd pushed the weight limits. And I hadn't accounted for the full distance needed to get airborne with that load - on a turf runway, in that heat, with a berm at the far end.

Lessons Learned:

Takeoff isn't complete until you're clear of everything.

I had become comfortable with the field - too comfortable. I knew the turf strip and had flown it before with similar loads. But on that day, the combination of weight, grass length, density altitude, and a hidden obstacle came together in a way I hadn't fully respected.

Fertilizer and fuel add weight, but confidence can add more.

Since that accident, I've changed how I evaluate strips. I walk the end if I haven't been there in the last few days. I verify obstructions - not just from memory or maps, but with my own eyes. And I don't assume a load that worked yesterday will perform the same today.

Every takeoff counts. And sometimes, it's not the air but the ground that gets you.

NOTES:

THE CROSSWIND LESSON
SCHWEIZER G-164B AG-CAT

It was a warm August afternoon, the kind where the skies are clear, the sun high, and everything about the day invites you to fly. I'd taken off from Fort Ripley, a short hop away, just finishing up another routine agricultural spray mission. Nothing about the flight itself was out of the ordinary - conditions were visual, winds manageable, and I was headed into familiar territory.

The destination was a private strip I'd used more times than I could count. Narrow, just twenty feet wide and just over half a mile long, but well maintained - enough to serve the kind of tailwheel work I'd spent decades doing. I'd logged over 8,600 hours in aircraft, the vast majority in this very model, a Schweizer G-164B. I knew its quirks, its limits, and how to coax it safely onto just about any surface.

Coming in from the north, I set up a straight-in approach. I'd been monitoring the wind; there was a steady breeze from the east, and I could feel it nudging me even on final. I added a few extra knots of airspeed to give myself better control authority in case the crosswind picked up. It's a trick you learn early when flying taildraggers - airspeed gives you options.

But my judgment was off.

Maybe it was habit, maybe a touch of complacency. Instead of touching down near the threshold, I floated longer than I should have and landed nearly two-thirds down the runway. Not ideal. The margin was already slim, but at the time, it didn't set off any alarms. I brought the power back, adjusted the prop pitch to low, and let the tailwheel settle in.

That's when it happened.

The second the tail touched the asphalt, the aircraft yawed - hard left. I felt it immediately, that sickening slip when you know the airplane's no longer tracking where you want it to. I was at the exact point where a break in the surrounding tree line funneled wind straight across the runway, catching the broad surfaces of the aircraft like a sail. The crosswind had intensified - more than expected - and I was now along for the ride.

The left main gear rolled off the edge of the narrow strip. I tried to salvage it - right rudder, braking, anything to bring her back - but it was too late. We skidded across the grass and into a plowed field. The soft earth caught the gear like a hook. The nose dropped hard, and in the blink of an eye, we flipped over.

Silence.

I hung there for a moment, suspended by the harness, staring at the ground where the sky should be. The dust settled around the cockpit, and I realized I was upside down, but unhurt. I unlatched, crawled out, and stood in the field, staring at the battered machine resting on its back. The vertical stabilizer was crushed. My pride, worse so.

Later, I reviewed everything. The weather reports confirmed what I already knew - the wind at a nearby airport was from 090 at 8 knots. A crosswind, but nothing extreme. But the strip I was using was aligned nearly north-south, and in that break between trees, that 8-knot wind became a blast across the runway with almost no warning. A trap I should've anticipated.

There were no mechanical failures. No surprises from the aircraft. This one was all on me.

Lessons Learned:

Crosswinds don't just affect landings - they test your judgment long before the tires touch the ground. Every airstrip has its own personality, shaped not just by its layout, but by the terrain around it. Gaps in tree lines, buildings, or hills can all funnel wind in unpredictable ways.

On paper, this was a routine landing. In practice, it was a reminder that familiarity can breed assumptions - and assumptions in aviation are a luxury we can't afford. When you're operating at the margins, especially in taildraggers on narrow strips, even a seemingly small decision - like a slightly long landing - can tip the balance.

Every landing starts long before the wheels touch down.

And *every* mistake teaches you... if you're lucky enough to walk away.

NOTES:

THE RUNWAY THAT BIT BACK
CESSNA A188B

It was a warm morning in August, and I was already deep into my fourth spray flight of the day. The first three had gone smoothly - light winds, cooperative air, and that early calm that makes everything feel easy. But by mid-morning, things had started to shift. The temperature was creeping up, the air felt heavier, and the wind had picked up slightly - nothing major, just a subtle shift that didn't raise alarms at the time.

I was flying a Cessna A188B, a rugged old agricultural workhorse I'd spent nearly two hundred hours in. I knew its strengths, and I knew its limits - or at least I thought I did.

The runway I was using was a soft, wet strip - not ideal, but common enough in ag flying. I lined up to the north as I had all morning, giving it full throttle and watching as the familiar hum of the Continental IO-520 filled the cockpit. Everything felt normal until I reached the far end of the strip.

That's when things started to go wrong.

The plane entered ground effect - close enough to the surface that it felt like flying, but without the lift to climb. I could feel it struggling to break free of the earth's grip.

I nudged the stick gently, trying to coax her into the air, but we weren't gaining altitude. The trees at the northwest end of the runway were coming fast, and I knew I had to act. I banked slightly right to avoid them, hoping to pick up some clean air or at least buy some space.

Instead, we descended.

The main gear hit the soybean field just beyond the runway. Before I could even finish processing that, we slammed into a drainage ditch. The nose pitched down violently, and the aircraft flipped forward.

Metal crumpled. Glass shattered. The world turned end over end in a blur of green and brown.

And then? Stillness.

I was upside down in the cockpit, the four-point harness holding me firmly as I took a mental inventory: bleeding, sore, but alive. I unlatched the harness, crawled out through the wreckage, and stood up in the quiet field, staring at what used to be an airplane.

Both wings were mangled. The fuselage twisted. The empennage bent into a cruel arc. She'd absorbed the crash like a warrior, giving me the chance to walk away.

Later, standing by the wreckage as adrenaline wore off, I played the whole takeoff over in my head. The weather report confirmed it: six knots out of 240. That meant a quartering tailwind from the left - barely noticeable, but enough to affect the climb on a short, soft runway. And that's exactly what I'd launched into.

No mechanical issues. No failures. Just a decision - a poor one, in hindsight.

I'd let the convenience of continuing operations override better judgment. I'd flown three successful missions that morning from that same strip, and I assumed the fourth would be no different. But conditions had changed. The runway had softened more with each pass. The wind had shifted just enough to matter. And I'd chosen to push forward anyway.

Lessons Learned:

Flying often demands boldness, but it rewards humility.

In agriculture, it's easy to develop tunnel vision - focused on coverage maps, chemical loads, and job completion. But no amount of efficiency justifies launching into marginal conditions.

A quartering tailwind and a soft strip might not look like a big threat on paper, but in the air, they combine into a deadly trap.

Every takeoff deserves a fresh assessment - especially when the wind shifts or the surface changes. There's a fine line between confidence and complacency, and I crossed it that day. It nearly cost me more than twisted metal.

If the runway's telling you no, listen. Because even when it doesn't speak loudly, it can still bite.

NOTES:

THE SNAP ON TOUCHDOWN
CESSNA A188A

It was a quiet evening in May, and I was wrapping up the final application run of the day over a field outside Canby, Minnesota. The sun was low, casting long shadows across the strip as I set up for landing. Everything had gone smoothly - solid passes, steady wind, and no surprises.

I brought the aircraft down gently, flaring just right. But the moment the tailwheel touched the runway, everything went sideways - literally.

The airplane jerked hard to the left and spun violently. I felt the gear collapse beneath me. The left wing dropped and scraped across the ground. The aircraft groaned, metal twisting against pavement, until we skidded to a stop facing the opposite direction.

I sat still for a moment, my hands frozen on the yoke. No smoke. No fire. I was unhurt. But the aircraft had taken a beating - the left main gear was destroyed, and the wing was bent and torn from the ground strike.

I radioed in and climbed out. The damage was clear: a textbook ground loop.

I'd flown thousands of hours, many of them in this exact model. So how did I end up spinning out on what felt like a perfect landing?

That answer came later.

Investigators examined the tailwheel spring cross tube - the heart of the tailwheel assembly. It had snapped during touchdown, causing the aircraft to lose directional control and spin. Metallurgical analysis revealed the part had likely been in service since the aircraft's manufacture in 1970. The design back then called for a thinner wall than current standards.

In 1985, the design was upgraded to a much thicker-walled part, nearly double the thickness. But the part on my airplane hadn't been replaced. Instead, it showed signs of sanding - likely to remove surface corrosion - which might have weakened it further. Just a thousandth of an inch here, a touch there, and suddenly a part that had served faithfully for decades was no longer up to the job.

The break wasn't from fatigue - it was overload. But the critical insight? That overload was only possible because I lost control during landing. And that loss of control triggered a sequence the airplane simply couldn't absorb.

Lessons Learned:

Age in aviation isn't just a number - it's a risk factor.

That part - steel, solid, and seemingly fine - had lasted more than 50 years. But service in agricultural aviation is brutal: chemicals, vibrations, and hard landings. Components that live under that stress can't be trusted forever, even if they pass the eye test.

But more importantly, this event reminded me that no part, no inspection, and no airworthiness certificate can replace the need for pilot vigilance. Taildraggers are unforgiving machines, especially when you let your guard down. Even with thousands of hours, I let habit take the place of precision.

It might've started with a weak part, but it ended with a moment of overconfidence.

Now, I check the parts that never make the maintenance list. The ones that "never fail." And every landing - no matter how routine - I treat like it's the one that might test everything I've got. Because on that day, the difference between walking away and rolling away was just a snap of steel I never saw coming.

<u>NOTES:</u>

THE TAILWIND TWIST
AIR TRACTOR AT-502

It was a calm-looking morning in August when I taxied out from Appleton. I had listened to the automated weather station - winds out of the east, about 8 knots. Runway 13 made sense, or so I thought. I even listened again, just to be sure. Same report. So I lined up, added ten degrees of flap, and prepared to roll.

But as I taxied, I glanced at the windsock. It didn't quite match the broadcast. It was pointing more from the north-northeast. Something about it made me pause - but I trusted the automation. Machines don't lie, right?

I brought the stick full back and slightly left into the wind, just like I always do. I advanced the throttle and started down the strip.

The takeoff roll was longer than usual.

At around 60 mph, I eased the stick forward to lift the tail - and the airplane suddenly snapped left. Hard. No warning. I stomped on the right rudder, trying to bring her back in line, but she was gone. The airplane veered off the runway and dropped into the drainage ditch alongside it.

The right gear collapsed. I felt the jolt run through the airframe.

Then both wings hit the ground, scraping and crumpling. It was over in seconds.

I shut everything down and climbed out, unhurt. But my airplane was another story - substantial damage to both wings, and the gear was a mess. And all I could think about was that damn windsock.

In the hours that followed, we pulled the weather data again. The station still showed 080° at 8 knots before and after the accident. But modeling later confirmed a surface wind closer to 061° at 4.5 knots - east-northeast. And that matched the sock. It was a quartering tailwind, and I'd let it fly right under my nose.

I hadn't lost the airplane because of a gust, or a system failure, or a malfunction. I lost it because I trusted the technology over my own eyes. The tailwind worked against me, and on a taildragger, that's all it takes to start a ground loop.

Lessons Learned:

Automation is a tool - not a replacement for situational awareness.

I listened to the AWOS, and it told me what I wanted to hear. But right in front of me was the windsock - an old-school, low-tech, wind-driven truth-teller - and I ignored it.

Quartering tailwinds are no joke, especially in an ag aircraft with a tailwheel. You give them a foothold, and they'll own your rollout before you even leave the ground.

Since that day, I make decisions based on all the inputs - not just the digital ones. I cross-check everything. AWOS, windsock, gut instinct. And when there's a disagreement, I side with the sock.

Because the airplane doesn't care what the broadcast said - it only responds to what the wind is actually doing. And if you don't stay ahead of that, it'll put you behind the ditch. Fast.

NOTES:

THE WAGON I DIDN'T SEE
UNSPECIFIED THRUSH

It was a clear afternoon when I climbed into the cockpit of a Thrush ag aircraft to conduct a demo flight for some potential buyers. I'd been given a map of several adjacent fields to perform spray passes. Everything seemed routine - familiar, even. I'd already made five clean passes without incident, flying low and steady, skimming the crops with practiced ease.

But on the final pass, my rhythm broke - and so did my propeller.

As I climbed out from that last run, I failed to account for a silage wagon parked near the edge of the field. The aircraft's left landing gear, propeller, and several spray components clipped the top of the wagon. Fortunately, I maintained control and returned to the departure airport without further issue. The aircraft was damaged, but I walked away unhurt.

Reflecting on the incident, I realized the root cause wasn't visibility, fatigue, or poor weather. It was my own mindset. A creeping sense of invulnerability had set in, the kind that whispers, "You've done this hundreds of times - you've got this." That complacency lulled me into a rhythm that didn't account for change or new threats - like that wagon at the edge of the field.

Lessons Learned:

Complacency doesn't arrive loudly - it creeps in quietly, disguised as routine. I'd flown clean passes all day, until one final climb-out clipped a silage wagon I hadn't accounted for. It wasn't bad weather or low visibility - it was the rhythm of familiarity that dulled my awareness.

What was clear five passes ago wasn't clear now.

This incident taught me that every pass deserves a fresh set of eyes. No matter how experienced, no matter how routine the mission, there's no substitute for humble vigilance. In ag flying, the moment you stop reassessing is the moment risk steps in.

NOTES:

WARNING LIGHT IGNORED
AIR TRACTOR AT-502B

I loaded 150 gallons of Jet A into the wing tanks that morning. It was going to be a busy run - multiple fields to cover, but nothing unusual. The air was calm, the light perfect. I was well into the swing of things by the time I loaded my third batch and lifted off again. That's when the light came on.

Subtle at first - the low fuel warning. A soft glow on the panel, something that usually gives you time to wrap up and return. According to the manual, that's when you stop everything: fly straight and level, monitor your fuel, and land as soon as practical.

But I didn't.

I figured I could finish the field. I was nearly done. The warning light had come on during a pass, and I thought I could squeeze in just one more turn. The aircraft was still flying strong. The gauges hadn't dipped too low. I'd been there before, and it always worked out. Until it didn't.

Midway through the next low pass, I noticed a change. The engine sounded different - softer. I nudged the throttle, and nothing happened. Power was fading fast.

I broke off from the spray run and tried to get some altitude, maybe find a place to set down. But the airspeed was bleeding away, and the engine was slipping into silence.

There wasn't time for anything but instinct.

I picked a bean field nearby, pushed the nose down to hold what little speed I had, and lined up. The touchdown was rough. We bounced, skidded across a small road, and slammed to a stop. The fuselage groaned. The right wing took a hit, and the aircraft was banged up, but I was okay.

A local came out to help. He had some diesel on hand, and together we added about 14 gallons total between the tanks - red-dyed road fuel, just to help get the investigation going. It was too late for anything but lessons.

There was nothing wrong with the airplane. No mechanical fault. No blocked lines. No leaks. The only problem was that the engine had run out of fuel. And the only reason it had run out was because I hadn't listened to the warning signs.

Lessons Learned:

Warning lights aren't suggestions. When the low fuel light came on, I had a choice: trust the manual and head back, or push ahead and hope for the best. I chose wrong.

Fuel exhaustion is entirely preventable. I knew better. But I overestimated the margins and underestimated the consequences.

In aerial application, the temptation to finish the job is constant. One more pass, one more turn, just another few minutes. But when you're this low, and this heavy, the line between "just enough" and "not enough" is razor-thin.

Next time that light comes on, I won't ignore it. I'll break off. I'll land. Because the field can wait. The client can wait. But when fuel runs dry at fifty feet, there are no do-overs.

NOTES:

PART 6 - COMMUNICATION

"Aviate, Navigate, Communicate."
　–Cornerstone aviation principle

A PRESIDENTIAL CLOSE CALL
UNSPECIFIED AGRICULTURAL HIGH WING AIRCRAFT

It was a routine morning mission - an agricultural dispersal flight over the San Jose area for a government agency. I was flying solo, operating under VFR, and everything checked out pre-flight. I ran the weather, reviewed NOTAMs via DUAT, and found nothing out of the ordinary. No restrictions. No warnings.

After departing Moffett Field (NUQ), I was handed over to San Jose Tower and directed to my designated drop zone for the dispersal of sterile Medflies. The task went off without a hitch. With my mission complete, I radioed NUQ Tower from 7 miles southeast, requesting permission to return and land. Clearance was granted: right downwind for Runway 14R. I followed the instructions and landed without issue.

Or so I thought.

Shortly after shutdown, I was approached on the ramp - not by airport ops, but by the Secret Service.

They wanted to know why I had flown so close to the President's party in Mountain View.

Caught completely off guard, I explained there had been no indication of restricted airspace.

No NOTAMs, no TFRs, and certainly no instruction from ATC to avoid the area. NUQ Tower had cleared me to land, just as they always had. If there had been an active restriction, I believed they would've rerouted me or denied entry. But everything appeared normal.

The irony? I was just doing bug control. And suddenly, I was in the crosshairs of a presidential protection perimeter.

Lessons Learned:

I followed the book. I checked NOTAMs, reviewed weather, and complied with ATC instructions - but still wound up face-to-face with the Secret Service.

Unbeknownst to me, the President was nearby, and despite no listed TFRs, I'd unknowingly flown within sensitive airspace. The takeaway? Standard checks don't always reveal the full picture, especially with VIP movements. TFRs can be delayed, unpublished, or dynamic for security reasons.

Since that flight, I always call Flight Service for an extra layer of assurance when flying near potential high-profile areas. Because even when you're just dropping bugs, the sky can hold invisible lines, and the consequences for crossing them can be anything but routine.

NOTES:

ABANDONED INFRASTRUCTURE HAZARDS
KIOWA WARRIOR

During an aerial application to a corn field, my helicopter's right skid became entangled in what appeared to be a well wire no longer in use. The wire did not break free; instead, it dragged the right side of the aircraft into the cornfield. The spray boom made contact with the corn, causing noticeable damage.

I immediately attempted to escape the drag on the aircraft, managing to gain some altitude. However, my egress routes were limited: I could either fly downwind and uphill, crosswind over tall trees, or headwind over 50-foot power lines. I chose to attempt the upwind route and fly over the power lines, but the helicopter remained entangled in the well wire, and I could not gain enough altitude to clear the lines. Unfortunately, I made contact with the power lines, which caused the chin glass of the aircraft to break and the power line to snap.

After the contact, I was able to recover control of the aircraft and safely land, suffering no further damage to the aircraft or injury to myself. A post-incident inspection revealed that a significant amount of well wire had wrapped around the right skid.

Lessons Learned:

This incident reinforced the importance of pre-mission reconnaissance. Not just from the air, but through direct communication with landowners.

Abandoned infrastructure like old well wires can be nearly invisible from above and pose serious hazards during low-level operations. I learned that relying solely on visual checks isn't enough.

Asking about previous field use, hidden obstacles, or buried infrastructure should be a routine part of our preparation. Additionally, in hindsight, I should have delayed my egress until reaching a safer location with more favorable conditions.

A lighter aircraft with more maneuvering room could have helped avoid contact with the power lines.

Sometimes, the safest move is the most deliberate one.

NOTES:

PARALLEL PATTERNS, HIDDEN DANGER
UNSPECIFIED AGRICULTURAL AIRCRAFT

Operating from a closed runway at an uncontrolled airfield isn't usually a problem - if everyone communicates and cooperates. But that's not always the case, as I learned over two consecutive mornings at the Newport Air Base in Arkansas.

I was flying agricultural missions north of the base, using the old, closed runway designated specifically for ag ops. Returning to reload, I set up a southeast approach, but spotted a helicopter parked on the strip. I adjusted, swinging around to land to the northwest instead.

Just as I aligned for that approach, I saw another aircraft - call it "Y" - coming in on a straight-in for Runway 18. The winds were out of the north at 10 knots, meaning he was downwind. I immediately climbed to 500 feet AGL and entered a wide pattern to keep our paths separate. He wisely went around, then followed a track that kept him well clear of me. I landed safely.

The next day, déjà vu struck. I returned from spraying fields east of the base and saw Y once again on final to Runway 18. I slowed down to give him time to land ahead of me. He instead chose to go around again. I maintained slow flight until he passed, then landed behind him.

At no point were we ever closer than 1,000 feet. But it was close enough for concern - and it happened twice.

Here's the bigger issue: The airport management has restricted agricultural ops to a closed runway that intersects general aviation traffic. It's not a setup built for safety. I've flown ag from controlled and uncontrolled airports for 14 years, and we've always integrated with the normal traffic pattern. But here, the split setup forces unnecessary conflicts.

I don't know if this was just coincidence or someone trying to get our ag operation out of the airfield. But when the same aircraft is involved in both close calls, and the rules are stacking against you, it's hard not to wonder.

Lessons Learned:

Using a closed runway for ag ops might seem efficient. But it only works if traffic flow is well-managed.

At Newport, our runway intersects active GA traffic with no coordination in place. Twice in two days, I narrowly avoided conflicts with another aircraft on straight-in approaches to Runway 18. Though no contact occurred, the setup invites risk.

Split operations without shared patterns or communication protocols are a recipe for near misses. Ag and GA can coexist, but only with clear expectations and smarter airport design. In the absence of ATC, structure matters more than ever.

Safety shouldn't hinge on mutual guessing.

<u>NOTES:</u>

THE SILENCE THAT SPOKE
UNSPECIFIED AGRICULTURAL LOW WING AIRCRAFT

I was crop dusting near Acadiana Regional Airport (ARA), operating from a private strip west of the field. The job required me to spray northeast of the airport, which meant repeatedly crossing the north end of the runway. First heading to the field, and then back again to reload.

ARA has a control tower, but it's lightly trafficked and operated by a private contractor. Each time I crossed the runway, I made radio calls to the tower. They responded consistently when I was heading eastbound - but not always when heading west. I figured the controller was just too busy with his malfunctioning radio and didn't bother to reply when traffic was low.

Over two days, I must have crossed the runway twenty times. On Sunday around noon, I made my usual call to cross. No answer again. Assuming it was business as usual, I went ahead. This time, however, the tower snapped back after I crossed: I'd just cut in front of IFR traffic on final.

I was surprised, and a bit shaken. When I asked the controller why he hadn't answered, he said he didn't hear my transmission.

Later, I made another call from nearly the same location and again got no response. But when I was just half a mile closer, the tower picked me up. That's when it hit me: there was a dead spot in my transmission zone.

Lessons Learned:

Over two days of routine ops, I mistook a lack of radio replies for a busy tower - not realizing I was transmitting from a dead zone. That assumption nearly put me in the path of IFR traffic on final.

The lesson? Never proceed without positive confirmation in controlled airspace. If the tower doesn't respond, don't just press on - troubleshoot, reposition, or hold until you know you're heard.

Communication gaps can be invisible until it's too late.

I now double-check signal strength and never rely on routine or habit to justify crossing active airspace. When it comes to radios, silence isn't harmless. It might be a warning.

NOTES:

PART 7 - WEATHER & WIND

"You take what the wind offers."
  -Anonymous ag pilot

INTO THE FOG OVER THE TREES
UNSPECIFIED AGRICULTURAL AIRCRAFT

I've been flying ag aircraft for more than 20 years - over 8,000 hours in the saddle - and I've come to know the quirks of Maryland's morning weather like the back of my hand. Our home strip, sits between a small river and a massive 200-acre sewage evaporation pond - an area I've dubbed "the foggiest place in the world."

It's not unusual for the area to be blanketed in morning fog, but typically by around 8:30 a.m., the sun burns through, leaving us with a workable margin of visibility. On this particular September morning, visibility was sitting at about 2 miles, with no clouds visible overhead. It looked clear enough to launch.

I loaded up a full tank of pesticide and rolled down the strip. As I cleared the treetops at the end of the runway, everything changed. Without warning, I ran into a wall of fog. Zero forward visibility.

With a heavy aircraft and precious little airspeed, I was behind the power curve and in real trouble. Instinct kicked in. I banked left, hard, doing everything I could to maintain ground contact. That turn brought me uncomfortably close to several homes nestled in the trees - closer than I'd ever want to be. But it allowed me to stay visual and eventually climb out northward, where the skies were clear.

Looking back, it was a close call. The entire area south of the runway was still socked in with fog. Had I delayed even a few minutes, I might've seen the full picture from the ground and made a different choice.

In this business, we operate near the limits all the time - often pushing minimum visibility and clearance standards out of necessity. But I'm reminded that familiarity with a field can be a double-edged sword. It can dull your caution.

Next time, I'll give myself a little more breathing room.

Lessons Learned:

This flight reminded me that familiarity can be blinding.

I knew the strip, the fog patterns, and still, I misjudged the margin.

Two miles visibility seemed safe - until it wasn't.

When I hit that sudden fog wall, instinct and luck saved me, not planning. It was too close.

The lesson? If there's even a hint of marginal conditions near confined terrain, wait. A few extra minutes can expose the hidden hazards the sky's holding back. Routine is never an excuse to rush.

In ag flying, where we already work on the edge, respecting weather - no matter how well we think we know it - isn't optional.

It's survival.

NOTES:

SNATCHED BY THE WIND
AIR TRACTOR AT-502B

I'd done this flight profile thousands of times - over 15,000 hours in this model alone. The morning was typical for south Texas in July: hot, a little hazy, and a stiff breeze working its way across the strip. I'd flown out of this small asphalt runway plenty, and the load wasn't out of the ordinary. I was confident in both the aircraft and the conditions.

As I eased the throttle forward, the big prop bit into the air and we started rolling. The takeoff roll was steady, and I rotated as usual. The aircraft lifted off smoothly. But just as I was clearing the end of the runway, it happened.

A gust.

Not just a puff, but a sudden hit - a wall of air that sapped the lift and stole my climb. One second I was clear, the next I was low, slow, and headed straight into the sugarcane.

The main gear clipped the tops of the 10-foot stalks. That was all it took. The airplane pitched forward and slammed into the field beyond. I landed hard on the right main gear, bounced once, and settled in nose-down. The right wing crumpled, the frame groaned, and everything came to a jarring halt.

I sat still for a moment, stunned, but unhurt. I'd been lucky. Very lucky.

There had been no mechanical issues. The aircraft was running fine. I didn't feel any torque pull, engine hesitation, or control surface lag. This wasn't about the machine - it was all in the air. A gust, poorly compensated for, at the worst possible time.

Later, I went back over every detail. My airspeed at liftoff. My pitch. My rudder input. I thought I had factored in the wind - but I hadn't respected the gust factor. I'd assumed the breeze was manageable and hadn't adjusted enough for the possibility of sudden variability.

Lessons Learned:

No matter your experience, the wind doesn't care.

It's easy, especially with thousands of hours in type, to get comfortable. To trust your habits. To believe that routine makes you immune to surprises. But a gust of wind at the wrong time doesn't care how many hours you've logged.

From that day on, I changed my approach. I check not just steady wind speeds but gust spread - and I plan for the worst. If there's a big gap between reported wind and gust, I wait. I recalculate. I change direction. Because when your wheels leave the ground, your margin for correction shrinks fast.

And when you're skimming treetops with a full load, there's no room to guess. You either account for the wind - or it will account for you.

NOTES:

TOO HOT A MARGIN
UNSPECIFIED AGRICULTURAL LOW WING AIRCRAFT

It was a scorcher on Maryland's Eastern Shore. The mercury pushed past 95°F, humidity hung in the air like a wet blanket, and the density altitude, according to the ESN AWOS, hovered around 2,400 feet. I was flying an agricultural spray mission about 15 nautical miles from base - loaded close to the aircraft's restricted gross weight.

The field was a tricky one: long rows flanked by congested areas and obstacles, demanding precision and discipline. I tackled the long rows first, hoping to burn off weight and give the aircraft a little more breathing room before maneuvering near tighter spots.

As I worked through the passes, I noticed the wind, which had offered some help earlier, dying down. With less cooling airflow and no real breeze, the aircraft's performance began to degrade noticeably - especially in the tight turns near the congested zones.

A few times, I had to pull harder than I was comfortable with to avoid overflying homes and barns. I made it barely. But I knew I was operating right at the edge of the aircraft's envelope.

Then it happened.

I was lining up for a final pass when I hit turbulence from my previous track. The sudden buffet caught me low and heavy.

To maintain control, I had no choice but to let the aircraft drift - right over the top of a congested area at approximately 200 feet AGL.

It wasn't much, but it was too much. Under normal density altitude conditions, the aircraft likely would've responded better. But with the heat, thin air, and reduced power margin, there was nothing left to give.

I was also running out of daylight and operating with minimum fuel - typical for weight-sensitive spraying operations under FAR 137. The pressure to finish before sunset played its part in narrowing my margins even further.

Thankfully, there was no incident, no damage. But it was a clear lesson:

Lessons Learned:

Heat changes everything.

That day, high density altitude robbed my aircraft of the power and performance I'd come to expect. Familiar turns became forced, safe margins narrowed, and turbulence pushed me into overflight I couldn't avoid.

The urgency to finish - low fuel, fading daylight - compounded the risks. Agricultural flying often demands operating near the edge, but this was a reminder that the edge shifts with temperature, humidity, and weight.

Don't rely on "normal" performance in abnormal conditions. Adjust your expectations, plan for degraded margins, and leave room to back out. Because when the heat rises, even a few feet of lost performance can separate a clean pass from a costly mistake.

NOTES:

WHIRLWIND DOWN
GRUMMAN G-164B AG CAT

It was May, and I was in the thick of the application season - flying fertilizer runs over rice fields in a remote stretch of Arkansas. The weather was clear, hot, and dry. I'd flown several passes already that day, each one routine. No turbulence, no issues. Just me, the airplane, and the endless patchwork of paddies below.

I lined up for another takeoff. Everything checked out. Engine sounded strong, airframe felt tight, and visibility was wide open. I was climbing out after liftoff, the aircraft heavy but stable, when the sky turned against me.

It happened fast - one moment I was gaining altitude, and the next, it was as if the airplane hit a pocket of air that wasn't there. The nose dropped and lift vanished.

I didn't see it coming. No dust, no tumbleweed spinning across the field. Just an invisible force: a dust devil.

About 8 to 10 feet wide, but impossible to spot until I was already in it. The moment I flew through it, the airplane rolled slightly and sank. My hands were already correcting, but it was like fighting gravity with a broken lever.

I tried to ride it out, but there wasn't enough altitude, or time.

The airplane dropped hard, clipped a dirt ditch, and flipped - nosed over and came to rest inverted. For a few seconds, everything was still. Then I realized I was upside down, still strapped in.

I released the harness and crawled out. The air reeked of dirt and fuel, but there was no fire. I was lucky.

The aircraft wasn't. The engine mount was wrecked, the wings and fuselage crushed, the tail snapped like a twig. From above, the damage looked like a toy tossed by a child mid-tantrum.

In the debrief, we found nothing wrong with the airplane. No mechanical faults. The engine hadn't skipped a beat. The airframe was solid. This wasn't about poor maintenance or bad decisions.

It was the air itself that turned on me.

Earlier in the day, I'd seen a few dust devils twist across the fields at a distance, harmless and fleeting. I made a mental note, but I never imagined one would align perfectly with my climb-out path. I didn't dump the hopper load. There wasn't time, and my focus was on keeping the aircraft flying. The truth is, once you're in it, a dust devil gives you no room to maneuver.

And here's the kicker - there's no formal training on this. No section in the FAA Weather Handbook that warns about dust devils. No advisory that tells you what to do when the sky spins at ground level and decides to knock you out of it.

Lessons Learned:

We train for engine failures, wire strikes, downdrafts, and bird strikes - but not for dust devils.

They're small, erratic, and mostly invisible. But they carry the power to flip a fully loaded ag plane and ruin your day in seconds. Especially right after takeoff, when your altitude margin is razor-thin.

I learned that day that even thousands of hours and a solid machine can't outfly invisible weather. You can't predict every variable. But you can respect the sky and never assume a clear path means a safe one.

Now, when I see those whirling ghosts on the horizon, I don't just make a mental note - I wait. I reassess. Because one thing's certain: no load is worth gambling with a force you can't see until it's too late.

Fly smart. Fly aware. And know that sometimes, the danger isn't mechanical - it's meteorological and silent.

NOTES:

BIBLIOGRAPHY

1. National Agricultural Aviation Association. (n.d.). Welcome to NAAA. Retrieved from https://www.agaviation.org
2. Aerial Application Association of Australia. (n.d.). About Us. Retrieved from https://www.aerialag.com.au
3. Aviation New Zealand. (n.d.). NZ Agricultural Aviation Association. Retrieved from https://www.aviationnz.co.nz/nzaaa
4. Canadian Aerial Applicators Association. (n.d.). Home. Retrieved from https://www.caaa.ca
5. Sindicato Nacional das Empresas de Aviação Agrícola. (n.d.). SINDAG – Aviação Agrícola Brasileira. Retrieved from https://www.sindag.org.br
6. Tylor Johnson Legacy Foundation. (n.d.). Safer Skies in Honor of Tylor. Retrieved from https://www.tylorjohnsonlegacyfund.com
7. Drone Federation of India. (n.d.). Home. Retrieved from https://www.dronefederation.in
8. DJI. (n.d.). Agricultural Drones. Retrieved from https://www.dji.com
9. Smithsonian Institution. (n.d.). Air & Space Magazine: Agricultural Aviation. Retrieved from https://www.airspacemag.com
10. Delta Farm Press. (n.d.). Ag Aviation Reports and Commentary. Retrieved from https://www.deltafarmpress.com
11. National Agricultural Aviation Association. (n.d.). PAASS Program Overview. Retrieved from https://www.agaviation.org/paass
12. Canadian Aerial Applicators Association. (n.d.). Operation S.A.F.E. Retrieved from https://www.caaa.ca
13. Air Tractor, Inc. (n.d.). Company History. Retrieved from https://www.airtractor.com/company/history
14. Thrush Aircraft, Inc. (n.d.). Aircraft. Retrieved from https://www.thrushaircraft.com
15. Federal Aviation Administration. (n.d.). Early History of Agricultural Aviation. Retrieved from https://www.faa.gov
16. Government of India. (2022). Drone Shakti Policy Brief. Retrieved from https://www.dronefederation.in
17. AgAir Update. (n.d.). Ag Aviation News and Safety Articles. Retrieved from https://www.agairupdate.com

18. Dakota Territory Air Museum. (n.d.). Home. Retrieved from
    https://dakotaterritoryairmuseum.com
19. Arsenal of Democracy. (n.d.). WWII Flyover Program. Retrieved from
    https://ww2flyover.org
20. Federal Aviation Administration. (n.d.). General Aviation Data &
    Trends. Retrieved from https://www.faa.gov
21. U.S. Census Bureau. (n.d.). Population and Demographic Trends.
    Retrieved from https://www.census.gov
22. CropLife. (2023). How Ag Pilots Cover 127 Million Acres a Year.
    Retrieved from https://www.croplife.com
23. Farm Progress. (n.d.). Role of Aerial Application in U.S. Agriculture.
    Retrieved from https://www.farmprogress.com
24. ARMoney & Politics. (n.d.). Turbine Takeoff: Life of an Ag Pilot.
    Retrieved from https://armoneyandpolitics.com
25. Mobility Foresights. (2023). Global Agricultural Aircraft Market
    Trends 2023–2030. Retrieved from https://mobilityforesights.com
26. Ag Aviation Magazine. (n.d.). Drone Tech and Aircraft Advancements.
    Retrieved from https://www.agaviationmagazine.org
27. Forbes. (n.d.). Aerial Agriculture in South America. Retrieved from
    https://www.forbes.com
28. SUAS News. (n.d.). Drones in African Agriculture. Retrieved from
    https://www.suasnews.com
29. Aerospace America (AIAA). (n.d.). Rise of UAVs in Agricultural
    Aviation. Retrieved from https://aerospaceamerica.aiaa.org
30. NASA Aviation Safety Reporting System (ASRS). (n.d.). Report a
    Safety Issue or Search Incidents. Retrieved from
    https://asrs.arc.nasa.gov
31. National Transportation Safety Board (NTSB). (n.d.). Case Analysis
    and Reporting Online Library (CAROL). Retrieved from
    https://carol.ntsb.gov

ABOUT THE AUTHOR

Fletcher McKenzie is a prominent figure in New Zealand's aviation community, renowned for his multifaceted contributions as a pilot, author, television producer, and aviation safety advocate. Holding a private pilot's license and two national flying competition titles, he has a rich background that includes roles as a tow pilot, paraglider, and parachutist.

His commitment to aviation safety is exemplified in his bestselling series, *Lessons from the Sky*, which compiles real-life incidents and near-misses to educate and promote safer flying practices among pilots. The series has achieved global reach, with over 20,000 copies sold. Building on that success came *From the Pilot's Seat*, a collection of narratives from 23 Kiwi pilots, showcasing diverse experiences from World War II missions to modern commercial flights.

Beyond writing, McKenzie co-created and produced the international television series *FlightPathTV*, which explores aviation stories and has aired on major networks worldwide. He also operates a global aircraft sales and parts supply business.

McKenzie's dedication to aviation extends to his service on various aviation trusts and executive committees, including Flying NZ (RNZAC). His extensive experience and unwavering passion continue to influence and inspire the aviation industry, both in New Zealand and around the world.

www.fletchermckenzie.com

www.ingramcontent.com/pod-product-compliance
Lightning Source LLC
Chambersburg PA
CBHW021658120626
46545CB00004B/1295